THEIR ARROWS
WILL DARKEN THE SUN

THEIR ARROWS
WILL DARKEN THE SUN

The Evolution and
Science of Ballistics

MARK DENNY

The Johns Hopkins University Press, BALTIMORE

© 2011 The Johns Hopkins University Press
All rights reserved. Published 2011
Printed in the United States of America on acid-free paper
9 8 7 6 5 4 3 2 1

The Johns Hopkins University Press
2715 North Charles Street
Baltimore, Maryland 21218-4363
www.press.jhu.edu

Library of Congress Cataloging-in-Publication Data

Denny, Mark, 1953–
    Their arrows will darken the sun : the evolution and science of ballistics /
Mark Denny.
        p. cm.
    Includes bibliographical references and index.
    ISBN-13: 978-0-8018-9856-3 (hardcover : alk. paper)
    ISBN-10: 0-8018-9856-0 (hardcover : alk. paper)
    ISBN-13: 978-0-8018-9857-0 (pbk. : alk. paper)
    ISBN-10: 0-8018-9857-9 (pbk. : alk. paper)
    1. Ballistics. 2. Trajectories (Mechanics). 3. Aerodynamics. I. Title.
    UF820.D46 2011
    623′.51—dc22   2010023459

A catalog record for this book is available from the British Library.

*Special discounts are available for bulk purchases of this book. For more information,
please contact Special Sales at 410-516-6936 or specialsales@press.jhu.edu.*

The Johns Hopkins University Press uses environmentally friendly book materials,
including recycled text paper that is composed of at least 30 percent post-
consumer waste, whenever possible. All of our book papers are acid-free, and our
jackets and covers are printed on paper with recycled content.

Contents

Acknowledgments

The process of publishing a book is one that takes time and the expertise of many people besides the author. For permission to reproduce photos from his collection, I am grateful to Bob Adams. For his excellent drawings of soldiers from classical antiquity, I thank Johnny Shumate. At the Johns Hopkins University Press, editor Trevor Lipscombe and art director Martha Sewall have worked their usual magic, while copyeditor Carolyn Moser has turned my turgid text into pristine prose—thank you all.

THEIR ARROWS
WILL DARKEN THE SUN

Introduction

Ballistic science has been important to humans for as long as they have hunted or waged war, but it is only over the past 250 years that the scientific discipline of ballistics has matured from a craft art. A deep understanding of modern ballistics theory requires specialized study over several years, and yet it holds a fascination for nonspecialists. You most likely would not have picked up a book explaining the science of thermodynamics, for example, though thermodynamics is probably more important to our daily lives—so why a book on ballistics? Maybe because of the popularity of hunting and target shooting. Ballistics and guns go together like ham and eggs, although, as we will see, ballistics applies to rocks, arrows, bombs, rockets, and flaming pianos as well as to bullets.

Why read a concise, semitechnical account of ballistics such as this one? Maybe because you're too busy earning a living to do a degree in ballistics but are left unsatisfied by some of the Sunday comics explanations of how it all works. You will not become a better shooter simply by reading this book: you know better than I do about the practicalities of your hobby, be it benchrest, bow hunting, or BBs. You will, however, gain a better understanding of ballistics science—of *why* ballistic trajectories are the way they are.

Ballistics has a long history—I will share some of it with you—because, of course, the development of firearms has greatly influenced the conduct and outcome of wars. Much of our understanding of ballistics principles came directly as a result of military development. Small arms and artillery weapons, their propellants and projectiles, plus rockets and ballistic missiles from arrows to ICBMs, have all evolved from simpler beginnings, over decades or centuries. Improvements led to longer ranges, which, as we will see, brought in new physics that required understanding (or, at least, accommodating) before evolution could continue. New understand-

ing led to yet longer ranges and greater accuracy, and so the cycle re-
peated. It continues to this day.

The figures for gun ownership give a false impression of the worldwide
interest in hunting and shooting. There are 192 million privately owned
firearms in the United States and almost none in England, though both
countries sport many hunters and shooters. Outside of the armed forces, it
is primarily hunters and target shooters who are interested in ballistics.
How many such folk are there? According to the National Shooting Sports
Foundation there are over 20 million active hunters in the United States
(including 3.3 million bow hunters) and over 23 million target shooters
(6.6 million of them are archers). They shoot with handguns, rifles, shotguns, air guns, longbows, and crossbows; they do it standing up, sitting
down, and lying prone. In England there are over 1,000 small-bore rifle
and pistol clubs; around the world the story is the same. Target shooting
with small arms was one of the founding sports of the modern Olympic
Games, beginning in 1896, and archery was only four years behind.

Keen interest in a subject naturally leads many people to want to understand it at a deeper level than they can glean from everyday experience.
Hence, sailors turn to aerodynamics to see how the wind fills their sails,
and target shooters turn to ballistics to learn the nuts and bolts of their
hobby (or profession—many military personnel need or want to know
about ballistics). To cater to the large public appetite for ballistics, there
are literally thousands of publications. These tend to fall into two broad
categories which I can encapsulate with two made-up book titles. The first
is *Splat! How to Build an Egg-Thrower in Your Garage!* and is generally aimed
at a younger readership. Such books tend to be light on technical details
and heavy on fun. At the other extreme you will find titles such as *Spin
Stabilization Data for .30-06 M2 Ball Ammunition: What Every M1903A4
Rotating-Bolt Springfield Rifle Enthusiast Needs to Know.* OK, so this is hyper-
bole, but you get the point: books in this second category tend to be
hypertechnical and very specialized. While full of technical details, however, they do not shed much light on the physical principles.

My book is aimed (an appropriate word) at the sparsely populated gulf
between these two extremes: I will share with you the *physics*, rather
than the engineering detail, that underlies ballistics (the *why* of ballistics,
which I mentioned a couple of paragraphs back). My purpose is to get
across to you the science of what is going on when you pull the trigger or
release the arrow. We will follow a projectile (arrow, bullet, artillery shell)
as it arcs through the air and see what effect it has upon a target.

You will not need a degree in engineering to follow the explanations, though I hope that professional ballisticians can learn from this book. Math analysis is included if you want it (but only if you want it) in a series of technical notes that follow the main text. My purpose is to get across the many physical principles that influence ballistics; to this end I will sacrifice details and perfect accuracy (this is why I call my book "semitechnical") because detail can cause confusion and cloud understanding. Instead I will use simple—but not too simple—models that convey the crux of an argument, without frills. If you crave the arcane details that will make you a better shooter, or if you want practical help building a toy ballista, then this is not the book for you. If you want to know why ballistics works the way it does,[1] then read on.

The scientific discipline of ballistics subdivides naturally into three parts, and my book adopts this convention. First there is the brief launch phase of a projectile. The object here is to provide the arrow, musket ball, or artillery round with as much energy as possible and send it in the desired direction. Second is the longer flight phase during which (until recently) the archer, shooter, or gunner has had no influence over the projectile—physics takes full control. Finally the terminal phase: our spear, sniper bullet, or mortar round hits something and penetrates, fragments, or explodes. In this book you will discover accessible explanations of all three ballistic phases, with or without math.

Math analysis is restricted to a series of technical notes at the end of the book because equations get in the way of readability. You may choose to omit the math analysis completely without missing out on any of the essential concepts, which require words to convey. If, on the other hand, you crave detail, there is enough of it in the technical notes for you to derive all the results shown in the physical models of ballistics phenomena that I construct.[2] So, whatever your technical requirement and whatever your interest in this engrossing subject, there is much in the pages to follow that will inform, stimulate, and amaze you.

1. Why does spin stabilize a bullet? Why do pistols have larger calibers than rifles? Why is gun barrel length so important? Why do bullets and shells drift when there is no wind? Why are longbows more efficient than guns?

2. The level of mathematical and physical sophistication in the technical notes varies from high school to undergraduate, including algebra, calculus, classical mechanics (Newtonian and Lagrangian), aerodynamics, elementary thermodynamics. . . . If math analysis doesn't light up your life, then the main text is designed to convey the concepts without math.

A final word about units and notation. As a physicist I naturally incline to the metric system, and so the detailed calculations that you will encounter in the technical notes, if you choose to do so, use dimensions of meters, kilograms, and seconds, plus their derivative units of newtons, joules, kilowatts, and so on. Much of the popular ballistics literature retains the old English system of measurement, and so in the main text I will occasionally use both systems, as in "a 32-pounder long gun from the Age of Sail had a maximum range of 2,600 yards (2,400 m)." Some modern firearm calibers are quoted in inches and others in millimeters; I will also freely mix these units, instead of converting. Common parlance refers to bullet *velocity* when, to a physicist, it should sometimes be *speed*. I will generally try to distinguish between these terms.

Part I

BANG!
Internal Ballistics

1 Before Gunpowder

Projectile weapons are as old as our species. Before gunpowder, weapons began and ended with the rock; Stone Age man would throw a small one by hand, whereas medieval man would launch a large one from a sophisticated counterpoise siege engine such as a trebuchet. Between these two extremes, other projectiles appeared: sling stones, arrows, throwing spears, and crossbow bolts. Ballistics enters into the trajectories of all these projectiles—and of musket balls, bullets, and shells—even before they leave the weapon that launched them. A projectile weapon, be it a bow, a trebuchet, or a Winchester rifle, is a machine designed to supply energy to its projectile. More specifically, it is designed to send the projectile at high speed in a very particular direction. *Internal ballistics* describes the process of generating the desired velocity;[1] *external ballistics* describes how a projectile flies through the air. In these first three chapters I will be describing internal ballistics and will begin here with the internal ballistics of pre-gunpowder weapons.

ONE HAND

Throwing a Rock

About as simple as it gets, you might suppose, and yet the biomechanics of rock throwing is far from trivial. The throwing arm is not a simple stiff lever that rotates to generate hand speed—though we will model it as such here. If we observe the throwing arm with more care, we see that it moves more like a whiplash, generating great speed at the thin end. Also, when you throw a rock (carefully chosen for shape and weight), you add speed

---

1. *Speed* describes how fast; *velocity* describes how fast and in what direction. People often refer to muzzle velocity when they mean muzzle speed.

by arching your body forward at just the right moment, and add force by pushing with your trailing leg.[2]

All this action (generated with barely a thought) serves to launch a rock perhaps 50 or 60 yards. We will learn later how maximum range relates to launch velocity: that is the subject of external ballistics. For now I will simply state that, to be thrown 50 or 60 yards, the rock must leave your hand with a speed around 75 feet per second (ft/s), angled upwards at about 45°. Suppose we were to approximate the action of a throwing arm by the rotation of a rigid rod about one end. We know that this is only a rough model (the more sophisticated model of Cross 2003 models the arm as a hinged rod), but it will suffice for my purposes. A stiff arm that is 2 feet long needs to rotate at a rate of five cycles per second (5 Hertz, or 5 Hz) to achieve such a range. In technical note 1, I show that a real arm (hinged at the elbow) can do better than this stiff arm. A hinged arm creates a whiplash effect that increases launch speed by about a third for the simple model of technical note 1—and probably by more for a real arm. This is why we throw rocks or baseballs with elbow initially bent.

In cricket, the pitcher (called a "bowler") is obliged to keep his throwing arm straight, reducing the ball speed. However, he gets some of this speed back because, unlike a baseball pitcher, he is allowed to run up to where he throws the ball, instead of standing still.

Throwing a Javelin

Olympic javelin throwers also run up to the launch point, to increase their distance. In this action, modern javelin throwers are imitating historical antecedents (fig. 1.1) who threw javelins or other throwing spears (such as the Zulus' *assegai*). A javelin is heavier than a sling stone, and so the point can cause damage to an enemy even if he is armored. Javelin throwers of classical history were often skirmishers who peppered the ranks of enemy heavy infantry, softening them up just before their own heavy infantry attacked. The Roman javelin (the *pilum*) had a characteristically long tip made of soft metal, with a barbed end. The metal would bend when it struck an enemy shield, so that the pilum could not be thrown back. If it penetrated the shield, it could not be easily removed, so the enemy soldier

2. The physiology of throwing is described by Chowdhary and Challis (2001); a physical model is provided by Cross (2003).

Figure 1.1. Ancient Greek infantryman, or *peltast*, with three javelins. Illustration courtesy of Johnny Shumate.

would be obliged to throw his shield away just as the Romans were about to attack.[3]

How much does running up to the launch point increase range? In figure 1.2 you can see that the maximum range increases by about 20% for a run-up speed of 10 ft/s (and 40% for 20 ft/s). Also shown in the figure is the optimum launch angle for the javelin; it increases from 45° when the javelin is thrown on the run. The graphs of figure 1.2 are derived in technical note 2; the range is obtained from launch speed without considering aerodynamic drag. For low-speed projectiles of limited range, such as

3. For Roman use of the pilum, see Webster (1980, p. 81) or DeVoto (1993, p. 132).

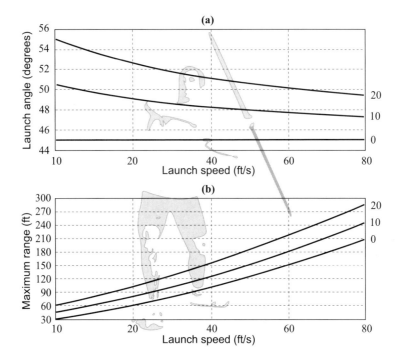

Figure 1.2. Javelin throw, for run-up speeds of 0, 10, and 20 ft/s. (a) Optimum launch angle vs. hand speed. (b) Maximum range vs. hand speed.

the javelin, this assumption is reasonable, but when we come to examine bullet trajectories, we will most certainly need to take drag into account.

The Sling

Our first projectile weapon, the sling, is an ancient and nearly universal tool, found all across classical Eurasia and also in Mesoamerica. Biblical peoples and ancient Greeks and Romans all used the sling as a military weapon. It dates from the Stone Age and was inexpensive and simple to make, though using a sling effectively takes a lot of training. In Roman times, Ballearic Islanders (in the western Mediterranean) were famous for their skills with the sling. More than one Roman historian records the story that boys on these islands were trained by withholding their food until they could hit it with a sling stone.[4] A Ballearic slinger is illustrated

---

4. Both Strabo and Vegetius (a fourth-century Roman general) make this claim. See, e.g., Vegetius, *De Re Militari*, book 1; for a convenient online source, see the bibliography.

in figure 1.3. The sling consists of two cords, traditionally made of wool or hemp, with a pouch in the middle to hold the sling stone. One cord end was fashioned into a loop, through which the middle finger of the throwing hand was placed (see fig. 1.3). The other loose cord end was knotted, so that it could be easily gripped between thumb and forefinger. The projectile is released by letting go of this knotted cord—at just the right moment. Skilled slingers were recruited for skirmishing with the enemy. They could send a projectile to a quite considerable distance—exceeding the range of ancient bows. The most common sling projectile was a rounded stone, but military use often led to specially manufactured clay or lead projectiles of biconical shape (like a pointed football); these flew farther than stone projectiles and did more damage.

A projectile in a sling pouch can be launched in one of several different ways. The simplest and most accurate, though with the shortest range, is the underarm shot. This is like a golf swing, with the sling replacing a club. A golf shot can send a ball 200–300 yards, but an underarm sling slot is shorter: the wrists cannot be used to power the sling through the bottom

Figure 1.3. A Ballearic slinger, in Roman times. Illustration courtesy of Johnny Shumate.

of a swing the way they do for a golf club. Also, the length of a sling that is used in underarm mode is quite short. Let us say that the length of arm plus sling is limited to a meter (just over a yard); with a maximum rotation rate of 5 Hz, as for the thrown rock, we find a sling projectile speed of about 100 ft/s (31 m/s) and a maximum range of 110 yards (100 m).

There is a sidearm delivery in which the sling is swung sideways—the same action as that of an Olympic hammer thrower. An overhead delivery is similar except that the sling can be rotated several times, like a lasso; here the sling length can be increased to perhaps 4 feet (say 1.2 m), increasing the range to 250 yards, or 225 m. Experienced slingers can do better than this, by taking advantage of the whiplash lever-arm effect that we saw in technical note 1 for the throw.

The current world record for a stone shot from a sling is over 440 yards. Longer ranges were claimed for slingers of the Old World classical civilizations. One problem with the sidearm or overhead delivery is that aiming is more difficult. Another problem for the military use of slings was that the sidearm and overhead deliveries require a lot more room, and so the number of slingers that can be brought into action at any one place is limited.

ANCIENT WARFARE 101

Many readers will have picked up a tolerable knowledge of modern warfare through their interest in ballistics, and if you are a professional soldier or veteran, you will have considerable knowledge on this subject. Fewer readers, I wager, will appreciate the role that ballistic weapons played in ancient warfare. So, here I will provide a broad outline of some of the features that ballistic weapons brought to the battlefields of ancient Greece and Rome—indeed to all battlefields of antiquity from prehistory to the dawn of firearms.

From a purely tactical point of view, the important factors that determined the outcome of battles were quality and nature of arms, troop density on the ground, and mobility.[5] The importance of warfare meant that a lot of resources and thought were put into getting it right, then as

---

5. Of course, there are many other considerations that determine the outcome of battles, such as soldiers' morale and determination, their physical condition, and the skill of their leaders. Here, though, I am concerned only with the influence of ballistic weapons and so will limit my discourse to factors relevant to these.

much as now, and so we find that warfare 2,500 years ago, say, was very highly developed in terms of army organization, troop specialization, and armament. Thus, heavy infantry (such as Roman legionaries and Greek hoplites) wore armor and carried close-quarter weapons such as swords as their main armament; they fought in densely packed formations that could maintain order while turning, or moving over rough ground. Light infantry wore no armor and carried projectile weapons such as javelins, slings, and bows. They were much more loosely organized, and because they carried less and were not in formation, they could move much faster, over rougher terrain.

Obviously, in close combat the light infantry would get pulverized by the heavies, but in reality the heavy infantry would rarely catch the light auxiliary soldier. Light infantry acted as skirmishers, spread out in front of the advancing enemy heavies, sending showers of projectiles at them from a safe distance, trying to break up their formation and soften them up so that friendly heavy infantry would prevail against them. As the enemy advanced, the skirmishers would withdraw behind their own heavy infantry units and let the opposing heavy infantry formations slug it out.

Cavalry was initially used to increase mobility about a battlefield. It was only gradually, over the course of centuries, that heavy cavalry was developed, taking up the role that tanks would adopt in modern warfare—rolling over the opposition, plowing through lighter units like bulls in a china shop. Light cavalry were often just light infantry on horseback: they could move faster but otherwise were no match for the foot soldiers. Light infantry skirmishers would beat light cavalry skirmishers most of the time—it was simply a matter of logistics. Foot soldiers could fire arrows farther than horse archers because until powerful composite bows (discussed below) became common, horse archer bows were weaker. Horse archers fired from a moving platform and needed to control their mounts, and so were usually less accurate. They carried less ammunition, and they presented bigger targets.

Heavy cavalry in dense formation were vulnerable to well-organized archers, as we will see, while skirmishing archers or slingers were easy prey for lighter cavalry units, who could run them down. Densely packed units of heavy infantry were similarly vulnerable to javelins, arrows, and sling stones but not to heavy cavalry (the horses had more sense than their riders and would refuse to charge into massed ranks of armored soldiers bristling with long spears).

So this is the mix of military units common throughout the battlefields of the ancient world: heavy and light cavalry and infantry; dense units equipped to fight in close order and diffuse units equipped to fight at a distance. Advancing technology (advancing more slowly in past centuries, to be sure, but advancing inexorably) influenced these battles by changing the delicate balance between different units. Chariots fell out of use, and then slings. Archery upgraded. Cavalry adapted. Firearms appeared and changed everything—but only slowly, as we will see in chapter 2.[6]

TWO HANDS
Staff Sling

A staff sling is a sling on a stick. As a weapon it lasted longer than the simple sling—well into the Middle Ages. It threw a heavier projectile than the simple sling could do, and farther. It also required two hands to operate. Imagine casting a heavy fishing pole; this is the action of a staff slinger. Why is the range greater than for a simple sling? It is because the lever distance—from shoulder to projectile—is increased. The arms hold a staff (which could exceed 6 feet in length) to one end of which a sling is attached. Note the progression of increasing lever distance. An arm with an effective length of perhaps 2 feet can throw a rock 60 yards; an arm-plus-sling of 4 or 5 feet effective length can throw upwards of 200 yards; an arm-plus-staff-plus-sling of perhaps 10 feet effective length can throw farther—how much farther we determine in technical note 3. The staff sling benefits from having three "hinges" (at the elbow, at the wrist, and at the point where the sling attaches to the staff) instead of two for the ordinary sling and one for the throwing arm. This increases the whip-lash effect. And with two hands swinging the staff, more power goes into the swing.

The staff sling mathematical model of technical note 3 assumes a staff length of 6½ feet (2 m) and calculates projectile launch speed for different sling lengths. The result is shown in figure 1.4. You can see that there is an optimum sling length of about 6 feet, say 90% of the staff length. This is true whatever acceleration the slinger applies to the staff. Also plotted is

---

6. To delve deeper into ancient warfare, see, e.g., Montgomery (1972); Sabin, van Wees, and Whitby (2007, chap. 13); and Sidnell (2007). The Greek *peltast* of fig. 1.1 was an innovation to combat skirmishers; peltasts were somewhere between light and heavy infantry. Note that the peltast in the illustration carries a sword as well as javelins.

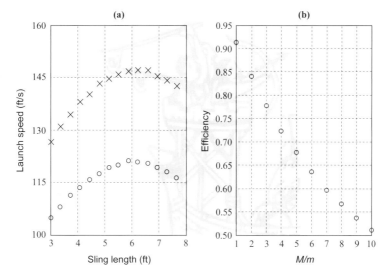

Figure 1.4. Staff sling launch speed and efficiency for a staff of length 6.5 feet (2 m). (a) Launch speed vs. sling length in feet, for two different angular acceleration rates of the staff: angular acceleration $\alpha = 20$ rad/s (o) and $\alpha = 30$ rad/s (x). In both cases (and for most values of α) the optimum sling length is about 6 feet. (b) Staff sling efficiency vs. mass ratio M/m, where M is staff mass and m is projectile mass.

the efficiency of the staff sling; note how efficiency falls as the ratio M/m of staff to projectile mass increases. This makes sense: it takes more energy to move a heavy staff than a light one. When the projectile is released, it takes away some of the energy that the slinger provided to the staff, but not all. Energy remaining in the staff (for example, kinetic energy due to staff movement) is not available to the projectile and so is wasted.

So what is the range of a staff sling? The angular accelerations of figure 1.4a are rather modest—they produce maximum rotation rates of 1½ Hz and 2 Hz—and yet they yield projectile ranges of 155 and 235 yards (140 m and 215 m), if we can neglect aerodynamic drag. A larger staff angular acceleration of $\alpha = 100$ radians/second (rad/s) produces a maximum staff rotation rate of 3½ Hz (less than that we generate when throwing a rock) and a launch speed of 274 ft/s (83 m/s). Ignoring drag, such a launch speed would send the projectile 770 yards (700 m) over level ground. However, we are now entering the region of high projectile speeds, and so we really cannot ignore drag. I will discuss drag in the chapters on external ballistics; for now, let us just say that a staff sling is capable of sending a projectile several hundred yards.

Longbow and Crossbow

The bow is almost as old and ubiquitous as the sling. The crossbow was known to the ancient Greeks and Chinese and was widely used in the Old World until well after the development of gunpowder weapons, as we will see. These two projectile weapons have certain advantages over the sling:

- They are more accurate.
- It is easier to repeat a trajectory with these weapons—to put two bow arrows or crossbow bolts into the same target.[7]
- Archers can be packed close together (an important military consideration).

The simplest type of bow is the *self bow*, which was made from a single piece of wood. A famous example of this type of bow is the English longbow (fig. 1.5). This type of bow was mass-produced in the fourteenth and fifteenth centuries as a military weapon: self bows are inexpensive, compared with the composite bows we will consider next, and are relatively easy to manufacture. Despite the simplicity of the longbow, considerable skill went into the construction. Wood was cut carefully from yew trees in winter, before the sap rose. The back of the bow (farthest away from the archer) consisted of elastic sapwood, which is strong under tension, while the belly (nearest the archer) consisted of heartwood, which is stronger under compression. The bow cross section was approximately D-shaped, with the back being flat and carefully cut along the grain. The wood was worked in slow stages over three or four years. Self bows do not have particularly long ranges. Even the powerful English longbows (which were over 6 ft long) attained ranges of only 175–240 yards (160–220 m). Their main tactical advantage was their high rate of fire—up to 12 arrows per minute.[8]

7. A slinger must exert effort throughout the entire internal ballistics phase, and so, to repeat a shot, he must reproduce exactly the motions of the previous shot. A bowman need only point his weapon in exactly the same direction as before, because once the bowstring is released, the weapon takes over. In other words, the slinger inputs energy throughout the internal ballistics phase of a shot, whereas the bowman inputs energy only *before* he aims his weapon.

8. The English longbow had an astonishingly heavy draw weight of around 100 lb. It required years of training to operate effectively, especially when the archer was part of a military formation. The fame of this longbow was not due to its

I don't include a technical note about bow internal ballistics because I have written one elsewhere;[9] instead, I will summarize some of the results here. The main surprise that emerges from all studies of bow dynamics is the efficiency of these machines. Indeed, an idealized bow (one which has a massless, inelastic bowstring and is not subject to friction or drag forces) is perfectly efficient. My model of internal ballistics for this idealized bow shows how energy is transferred entirely to the arrow just before it is released from the bowstring (see fig. 1.6a). Real bowstrings have mass, and this mass reduces efficiency.[10] The critical parameter is the ratio of bowstring mass to arrow mass. A modern bow with a Dacron bowstring weighing 7 g (¼ oz) and an arrow weighing 25 g (just under 1 oz) has an efficiency of about 90%. This means that 90% of the energy that the archer invests into drawing back the bowstring is transferred to the arrow. A medieval bow would have had a heavier arrow (English longbow arrows were long, and some were equipped with an armor-piercing arrowhead) and a much heavier string (made typically of plaited hemp), and its efficiency would be lower—perhaps 70%–80%.

A second feature that emerges from my model (and others) is the rapidity with which the arrow accelerates (see fig. 1.6b). In 17 milliseconds the bow accelerates its arrow from zero speed to about 200 ft/s (say 60 m/s). For the English longbow of figure 1.5 this short period of acceleration

<hr />

range or accuracy (it was used militarily against dense formations) but instead was due to the training of its archers and the logistics of their deployment. At the Battle of Crécy in 1346 some 7,000 English archers decimated charging French heavy cavalry—previously considered invincible—by firing half a million arrows at them. At their maximum rate of fire, these archers would have dispatched their missiles at the rate of 2 tons per minute. See Fowler (1967, p. 108), Bradbury (1992), and Hardy (1993) for details of the English longbow and its use on medieval battlefields.

9. See Denny (2003). Those readers who are following the math may refer to my Web site (Denny 2010), where a copy of this paper is available. The math model I construct is approximate; bow dynamics is complex, not least because it involves beam theory, so my model involved simplifications. For the mathematically inclined reader, references in my paper will direct you to articles that present the detailed, full-blown calculations.

10. When the arrow leaves the bow, the string vibrates; and if the bowstring has mass, this vibration consumes energy.

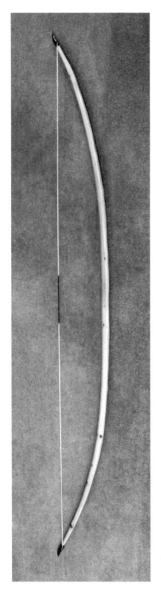

Figure 1.5. English longbow. A simple self bow—one made from a single piece of wood—this example is 6.5 feet long and has a draw weight of 105 pounds and a draw distance of 32 inches. Photo by James Cram.

means that the bow is imparting energy to the arrow at an average rate of about 10 kW.[11] For medieval technology, this is astounding.

Composite bows were more sophisticated than self bows. Composite bows were made of wood, horn, and sinew, glued into place. The wood was cut along the grain for tensile strength. The horn, on the belly side of the bow, was elastic but very strong in compression. Sinew, from the leg or neck tendons of cattle, was attached to the back side of the bow because it is strong in tension. Such bows were the result of a great deal of historical trial and error, presumably over a considerable period (they evolved surprisingly early—certainly they were common in Eurasia by the first millennium BC). Many cultures developed composite bows independently. Thus, Inuit peoples in the Canadian Arctic developed bows with sinew strung under tension, rather than glued, along the back side of the bow. More southerly American Indians also made use of sinew to prestress their short bows, either by the Inuit method or using glue.

The composite bow offered two advantages over the self bow: it was more powerful than a self bow of the same size, and it was short enough to be used by cavalry. Most of the nomadic tribes that periodically swept across Asia—wreaking havoc in China and Europe over a thousand

11. The energy stored in the bow is approximately ½ *Fd*, where *F* is draw force and *d* is draw distance. The bow power is determined by dividing this energy by the time it takes to fire the bow. For the numbers provided in the caption to fig. 1.5, we obtain the figure of 10 kW.

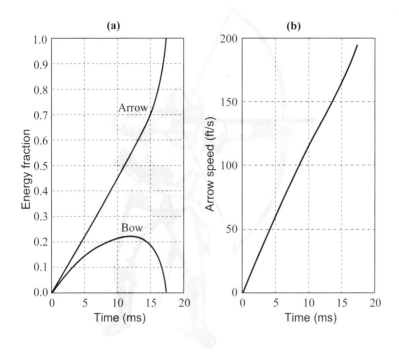

Figure 1.6. Internal ballistics for an idealized longbow. (a) Fraction of stored energy taken by the bow and the arrow during the launch phase. The bow gives up its energy entirely to the arrow. (b) The arrow is accelerated by the bowstring until they part after 17 milliseconds.

years from the fourth century AD—were armed principally with short composite bows. The main disadvantage of this bow was its complex and expensive construction.[12]

The high point of ancient bow construction is often considered to be epitomized by the Persian and Turkish composite bows. These bows became strongly *recurved* during the construction process. An unstrung recurved bow curves away from the archer, sometimes to the extent that the bow tips meet or cross. This means that the braced (strung) bow is under significant tension, increasing the draw weight and hence the power of the bow. The range of composite recurved bows is surprising—but that is a topic for another chapter. In figure 1.7 you can see why recurved bows are more efficient than bows that are not recurved. As the arrow is released,

12. For more on the history and technology of bows, see Denny (2007, chap. 1).

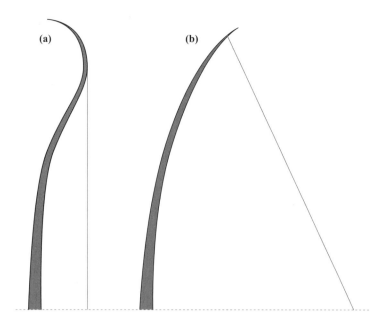

Figure 1.7. Recurve bows are more efficient than other bows because the effective bowstring length of the braced bow is reduced. (A bow is *braced* when it is strung but not drawn.)

the bowstring straightens. Because part of the string is now in contact with the bow, the string cannot vibrate when the arrow flies off, so the amount of energy that is wasted by string vibration is reduced.

Crossbows are extremely powerful short bows attached to a stock—in some ways the forerunner of muskets, as we will see. The bow itself is very stiff and difficult to span (to draw) because it is thick and made of composite material such as horn and wood, or of steel (even examples from medieval Europe are of steel). Crossbows from classical antiquity could be spanned by hand. The bowman could pull on the string with both hands, while the stock was held against his belly.[13] Medieval crossbows were more powerful. They could not be spanned by hand but required some sort of mechanical device to "cock" (to anticipate a term from musketry) the bow. In figure 1.8 you can see a simple λ-shaped lever that spans the bow. More powerful bows required a windlass device consisting of a ratchet and

13. The crossbows of ancient Greece were known as *gastraphetes* (belly-shooters) because of the way they were spanned. Crossbows of classical antiquity are discussed by Landels (1978, chap. 5).

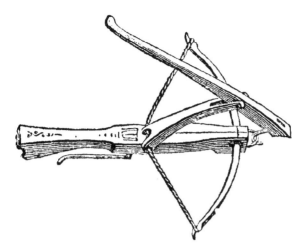

Figure 1.8. A crossbow, with a lever device for spanning. Note the trigger mechanism. The bolt is placed along a longitudinal groove in the stock.

pulleys. All this power released a *bolt* (as crossbow arrows are known) that could travel perhaps 380–440 yards (350–400 m). As you can see from figure 1.8, the bowstring was released by a trigger mechanism—again, a forerunner of the trigger used for muskets.

The advantage of crossbows over longbows was that the bowman needed relatively little training and strength. The disadvantage was the slow rate of fire—perhaps one or two bolts per minute. For certain applications, such as hunting, rate of fire was unimportant; it was much more useful to have a weapon that could be "cocked" and therefore ready to fire in an instant. The internal ballistics of crossbows is the same as that of longbows, though the parameters are different. From the perspective of the ballistics historian, crossbows are significant because they are closely analogous to—and indeed in many ways anticipated the early use of— gunpowder weapons. Crossbows overlapped with early hand-held gunpowder weapons and provided the standard against which these new weapons were measured.

Archer's Paradox

There is an interesting puzzle associated with longbow internal ballistics. If you are an archer, you will know about the *archer's paradox*, but few other people have heard of it, let alone know its resolution. I wrote about it a few years ago, but since then some impressive high-speed photographic evidence has been released that dramatically confirms the technical explanation.

Picture a longbow viewed from above. It is drawn and the arrow is about to be released. Hit the pause button; picture the arrow geometry in your mind. The back end of the arrow has a groove (the *nock*) that holds the bowstring in place. The arrow shaft lies alongside the bow handle. Because of bow handle thickness, the arrow is not pointing exactly at the target but instead is offset, pointing a little to the left (if the arrow shaft passes to the left of the handle). Now resume the motion, as our archer releases the bowstring. The string rapidly approaches the bow handle, and so the arrow accelerates but also (think about it) increases the offset angle. Surely, then, at release the arrow will move away to the left of the target direction, due to deflection by the bow handle? This makes sense, but it does not happen —real arrows are released in the target direction. Hence the paradox.

The explanation is strange. A perfectly stiff longbow arrow may indeed follow a trajectory that deflects from the target direction, but real arrow shafts are far from being stiff. What happens is this: rapid acceleration by the bowstring causes a wooden arrow shaft to buckle. That is, the shaft bends longitudinally under pressure from the string, and the arrow head realigns with the line of sight direction due to this bending. Once past the handle, the arrow then bends back the other way: it vibrates in a horizontal plane. The vibration is initiated by the sudden impulse provided by the bowstring; its frequency depends upon arrow shaft characteristics. You can see that it is important, if the arrow is not to be deflected by the bow handle, that the vibration must not be too fast or too slow, because then the arrow fletching (the feathers) would strike the handle. At just the right frequency, the arrow bends around the handle, barely touching it after the initial release. This is why it is important for archers to match the bow and arrow: as with firearms, the ammunition and the weapon must be designed for each other.

If my explanation of the archer's paradox has confused you (the physics is far from trivial), I urge you to look at two recent online videos that film arrow release in very slow motion. In one video, the camera is positioned behind the archer; in the other it is in front. Both show quite clearly how the arrow vibrates horizontally in flight—it more resembles a swimming eel than a rigid rod—and how it bends around the bow handle and follows the line of sight direction.[14]

14. Two YouTube videos listed in the bibliography (2007c and 2008b) demonstrate the archer's paradox more eloquently than any math analysis can. For further technical details, see Denny (2005, 2007).

MANY HANDS

Siege engines were the heavy artillery of olden days. Varied in form and size, these machines dominated siege warfare for two thousand years. I will consider two very different examples here. The *onager* was a throwing weapon of classical antiquity, familiar to both the Greeks and Romans, that was powered by the torsion of twisted rope or sinew. The *trebuchet* was a much larger siege engine of the medieval period, widely used across Europe and the Middle East, that was powered by gravity. In a way that was analogous to the competition between crossbows and early hand guns, the trebuchet was an important competitor of fourteenth- and fifteenth-century European cannon and set the standard for these early artillery weapons.

Onager

The onager was powered by twisted sinew or rope (or even human hair). (See fig. 1.9 for a Roman example.) The throwing arm was ratcheted back into a cocked position, a projectile (usually a round stone) was placed in position, and the arm was released by a trigger mechanism. The arm would then fly forward, to be stopped abruptly by a padded beam or post, at which point the projectile was launched. The abrupt stopping of the throwing arm would cause the rear end of the engine to kick upwards—hence the name (an onager is a type of wild ass). Analysis of the onager motion is simple compared with that of the bow or the trebuchet, and leads to the following expression for the efficiency of this classical siege engine: $\epsilon = m/(m + \frac{1}{3}M)$, where m is projectile mass and M is throwing arm mass.[15] Typically, the throwing arm might have been twice the weight of the projectile it threw, in which case the onager efficiency would be $\epsilon = 0.6$, i.e., 60% of the energy that was required to cock the throwing arm would be transferred to the projectile.

The onager was obsolete by the Middle Ages for a number of reasons:

- The sinew or rope spring that provided power had a limited lifetime and was susceptible to degradation in damp weather.
- The spring could be made to work only for small and medium-sized engines; large projectiles required a different mechanism.

15. My onager paper, available on my Web site (Denny 2010), contains a derivation of the equation for efficiency.

Figure 1.9. A Roman onager. Here, the projectile is placed in a short sling; in other designs the sling is replaced by a spoon-shaped depression at the end of the throwing arm.

- The kick stressed the machine, limiting its lifetime.
- Most important of all, the recoil kick meant that the position of the onager had to be reset after every shot. This reduced accuracy (especially important in that siege engines were often required to aim at the same section of wall or tower with many consecutive shots) and also reduced the rate of fire.

Trebuchet

All four of these problems were solved in the Middle Ages with the trebuchet (illustrated in fig. 1.10). The trebuchet belongs to a class of siege engines known as *counterpoise* engines because the power comes from a heavy counterweight at the short end of a hinged throwing arm, as shown in the figure. The throwing arm acted like a lever; the throwing end of the arm moved four or five times faster than the counterweight end because the hinge was four or five times closer to the counterweight. In addition, there was a long sling attached to the throwing arm, which effectively increased the arm length and provided the whiplash effect that we encountered earlier. The mechanism was smooth; this staff sling on steroids gently released its projectile and then continued its swing, like a golf club swinging through the hit.

Trebuchets could be built big, and were: an early version of the arms race in medieval Europe and the Middle East led to the growth of these engines. Castle walls were built stronger and higher, so trebuchets were made bigger to throw larger rocks to knock them down, so castle walls were made stronger to resist these projectiles, and so on. By the time cannons were well enough developed to put a stop to the proceedings (by

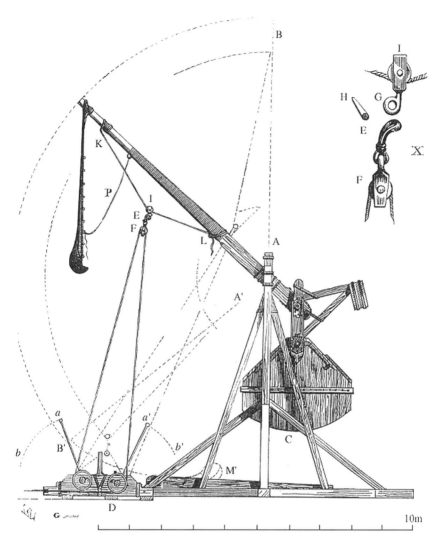

Figure 1.10. A trebuchet—the heavy artillery of the Middle Ages. The throwing arm has a sling at one end, containing the projectile, and a counterweight at the other, providing power. Some of these machines were massive, with counterweights up to 20 tons. This illustration is from *Dictionaire Raisonné de l'Architecture Française du XIe au XVe Siècle* (1854–68).

forcing a radical redesign of castles, as we will see in the next chapter), both castles and trebuchets were very large. The largest trebuchets launched heavy projectiles (usually rocks, though sometimes rotting
corpses to spread disease among the besieged) and were powered by
counterweights of up to 20 tons.

The internal ballistics of trebuchets are quite complicated, so here I will
simply summarize the results of my analysis.[16] A large trebuchet can throw
a 220-pound (100-kg) projectile a distance of 275 yards (250 m). Lighter
projectiles can be sent farther; heavier projectiles less far. (Historical
records claim some large trebuchets could fire projectiles that weighed
over a ton.) The efficiency of these machines exceeded 50%. Modern re-
constructions are able to group their shots closely—within a few yards at
ranges of a couple of hundred yards. The whiplash effect works well only
for certain geometries; crucial parameters include the ratio of short to
long lever arm lengths (i.e., the position of the throwing arm hinge—see
fig. 1.10), the ratio of long throwing arm length to sling length, the relative
values of projectile mass, throwing arm and counterweight masses, and
the initial angle of the throwing arm prior to trigger release. Medieval
siege engineers did not have the theoretical understanding that we have,
of course, and yet they managed by empirical means to produce near-
optimum engine designs.

Pre-gunpowder mechanical projectile weapons made use of the lever principle and of the whiplash effect to maximize the launch speed of projectiles. Their power came from human muscles (as in the case of thrown
rocks or slings), from the release of stored elastic energy (bows, ona-
gers), or from stored gravitational energy (trebuchets). Crossbows and
trebuchets provided the performance levels against which early gunpowder weapons would be measured.

16. See Denny (2005) for a technical analysis. This paper is available online at
Denny (2010). A less technical account of trebuchet dynamics, plus historical notes
of the evolution and development of these fearsome machines, can be found in
Chevedden et al. (1995) and in Denny (2007, chap. 3). Many trebuchets are still
being built, for educational purposes in university engineering departments or
simply for fun. See www.youtube.com/watch?v=-wVADKznOhY for a video of a
large modern trebuchet throwing a flaming piano and a small car.

2 Gunpowder Weapons

In chapter 1 we saw how the potential-energy weapons of bygone ages converted stored gravitational energy or the elastic energy of stretched sinews into the kinetic energy of a projectile. Over millennia, these weapons evolved and became impressive machines—impressive for their clever design as well as for their destructive power. Beginning in the Middle Ages, however, a new kind of stored energy began to be used. The chemical energy stored in gunpowder came to dominate projectile weapons. The process was a slow one, as we will see, because new technology had to be developed, and an understanding (initially empirical; theoretical knowledge arrived later) of the propellant properties of gunpowder needed to be acquired before gunpowder weapons changed the course of battles and therefore of history. But I am getting ahead of myself; first we need to see where gunpowder came from and how it led to—and powered—weapons of war for 600 years.

BLACK POWDER: HISTORY AND CHEMISTRY

The word *gunpowder* is a historical hangover that badly describes this important substance. Originally a powder that was not intended for use in guns, it became a grain (not a powder) that was so intended. In between there existed a brief period of time during which, yes indeed, the substance in powdered form was used as a propellant in guns—and the name stuck. Nowadays the word *gunpowder* applies to other gun propellants, and the original propellant that was called gunpowder has been renamed, with only slightly greater accuracy, as "black powder."

Confused? Examining the history and chemistry of black powder will hardly help dispel the smoke, but it is an interesting story, here encapsulated briefly.

Fire Medicine, Fireworks, and Warfare

The invention of black powder is commonly attributed to Chinese alchemists of the eighth or ninth century, and this may indeed have been the case, although at least one source claims that black powder, or something very like it, was known centuries earlier in India. Wherever the truth lies (and it is possible that both claims are true—black powder may have been invented more than once), it seems that the stuff was initially used for purposes other than war. Those early alchemists utilized *huo yao* ("fire drug") as a medicine. The Chinese also used black powder for fireworks. Today, more than one thousand years later, this peaceful application of the world's first explosive is once again the main reason it is made.[1]

It did not take long, however, for the Chinese to learn that black powder had many applications in warfare. Its military use dates from the tenth century—indeed, black powder was the world's only explosive until fulminates were developed in 1628, and it was the only gun propellant until the late nineteenth century.[2] Chinese soldiers lobbed incendiary bombs from siege engines into enemy castles. By the twelfth century Chinese warships used black powder in rockets that propelled fire arrows *en masse* against enemy vessels. The first Chinese cannon—more accurately, their earliest cannon that has survived down to the present day—dates from 1332, and cannon were used by both sides during a 1359 battle against the Mongols near Hangzhou.

The Mongols spread the use of black powder across Asia to the Arabs of the Middle East, and from there it quickly found a home in Europe. Initially, Europeans exploited black powder for bombs and mines but soon concentrated on cannons. Black powder had a greater influence on European history than on Chinese or Indian history because Europeans learned quite early to apply the stuff primarily as a propellant rather than as an explosive. Black powder works better as a propellant than it does as an explosive although, confusingly, it is generally considered to be an explosive. (I will unpack the properties of black powder and dispel this confusion soon enough.)

Here is the popular story about how Europe learned of the propellant

---

1. Black powder is also used today for ignition charges, primers, fuzes, and blank-fire charges in military ammunition. As late as 1968 the U.S. Army was using four million pounds of the stuff annually.

2. Fulminates are friction-sensitive explosives that were developed for use in percussion caps in the early nineteenth century.

property of black powder. In 1330 a German monk, Berthold Schwartz of Fribourg, was mixing black powder in a mortar (as in "mortar and pestle" —not the artillery weapon). Obliged to leave this task, he covered the mortar with a large rock to prevent contamination. The powder unexpectedly exploded, blowing the rock into the air. Pure myth. The good monk probably did not exist, although the date is about right for the first military use of black powder in Europe.[3]

Composition and Properties

Before I summarize the long evolution of gunpowder weapons in Europe, we need to understand more about the chemical and physical properties of black powder. The chemical composition is a mixture of three simple powders: 75% saltpeter, 15% charcoal, and 10% sulfur. Saltpeter, or potassium nitrate, occurs naturally as the mineral niter; an abundance of niter in China may explain the early use of black powder in that part of the world.[4] Charcoal is one form of the element carbon, readily created by the incomplete combustion of wood, and sulfur is another naturally occurring element. The proportions listed above are appropriate for propellant gunpowder; blasting powder (used in mining) contains less saltpeter and more sulfur.

So what does this mixture of powders do? When ignited black powder undergoes the following basic chemical reaction: $2KNO_3 + S + 3C \rightarrow 3CO_2 + N_2 + K_2S$. In words: potassium nitrate plus sulfur plus carbon yields carbon dioxide gas plus nitrogen gas plus a solid, potassium sulfide. In fact, the reaction is variable and more complicated than indicated here; ignition produces other gases, such as carbon monoxide and nitrogen oxides, as well as other solid potassium compounds. The main point is that, once black powder is ignited, the reaction proceeds quickly, and some 44% of the original mixture is converted into gases, while 56% turns

3. For the origin and development of gunpowder and gunpowder weapons, the reader can do no better than consult Hall (1997). See also Britannica (1998, s.v. "Ballistics"), Buchanan (2006), and Kelly (2005).

4. Europeans initially obtained niter from fermented urine. Later, they found it in horses' stables (from mortar that had been exposed to animal dung) and also from certain plants. The process of extracting saltpeter from these sources was initially difficult, so Europeans imported most of their gunpowder, at very high cost. This situation lasted until the fifteenth century and slowed the development of gunpowder weapons in Europe.

Figure 2.1. Napoleon's "fog of war," black powder generated a lot of solid by-products when burned, much of which emerged as smoke. Here, a musket from the late eighteenth or early nineteenth century generates smoke from both the primer charge in the flash pan and the main charge in the barrel. U.S. Navy photo.

into inert solids (usually appearing as white smoke—as in fig. 2.1—or as a residue deposited on the inside of gun barrels).

For loose powder in open air, the volume of gas produced is about three hundred times the original volume of powder; for compacted powder in a confined space, the volume increase is much higher. This is what makes black powder a good propellant: a small volume of solid material becomes gaseous within a few milliseconds, and this gas expands rapidly to its natural volume. Black powder can be made from saltpeter and charcoal alone, but it is stronger if sulfur is present. Sulfur does not directly contribute to the explosive force but, by uniting with potassium, generates a lot of heat—it raises the temperature over 2,000°C—which in turn increases pressure. (Interestingly, charcoal burning on its own will release more energy than black powder of the same weight, but this energy is released much more slowly.)

Unlike more modern nitrocellulose propellants, the combustion products of black powder are not well defined—and indeed the whole process of black powder ignition and combustion was not understood even in the middle of the twentieth century, long after black powder ceased to be the preeminent military propellant.[5]

The basic formula for black powder was first recorded in China in 1044, and in Europe around the year 1300. The earliest European illustration of a gun operated with black powder is in a manuscript of 1326 by the English scholar Walter de Milemete. It is known that within a few years the powder was being manufactured in England and in what is now

5. See Blackwood and Bowden (1952).

Germany. Cannons powered by black powder were used at the Battle of Crécy in 1346 (Edward III of England was known to be keen on the new technology). This is ironic (and a sign of things to come) because, as we saw in chapter 1, Crécy is remembered as a battle which showed the power of the English longbow—serried ranks of longbowmen sent hundreds of thousands of arrows into charging French cavalry, mowing them down. The use of cannons in these early days was largely for morale; they made a loud bang and produced a lot of smoke but were wildly inaccurate and may have posed more danger to their crews than to the enemy in front of them.

At first, black powder really was a powder—fine as dust—but it was found that such powder caused the bore of cannons to become gummed up after a few discharges. Over time, gunners learned that granulated (*corned*) black powder not only left less deposit inside the barrels of their guns but also increased the range of their projectiles. In fact, over the centuries it came to be realized that the size of grain had a huge influence on the muzzle speed of the projectile. Each grain burns inward from the surface, and so larger grains burn more slowly because they have less surface area than the same weight of small grains or powder. Gunners learned that they could control the rate at which granulated black powder generates gas by varying the size of grains (see technical note 4).[6] Perhaps counterintuitively, this slowing down of the combustion process leads to *increased* projectile velocity, for reasons that I will demonstrate in this chapter.

The expansion of gas from black powder is supersonic (though much slower than modern explosives, which generate significantly more shock waves). If this were not the case, black powder charges would have burst the gun barrels of early cannons. A key feature of propellants is that they release gas more slowly—over a period of milliseconds—than do true explosives, which detonate within microseconds. Modern propellants burn at a rate that changes with pressure, whereas the combustion rate of black powder is not very sensitive to pressure. This is why black powder is technically classified as an explosive, even though its *brisance*—the rate at which it generates gas—is low enough for it to function effectively as a propellant. The interplay between the variable rate of ignition of corned

6. *Corning* black powder—converting it from a powder to a grain—may originally have been done to prevent the powder from spoiling in wet weather. If so, the improvement in propellant quality was a serendipitous result. See Hall (1997).

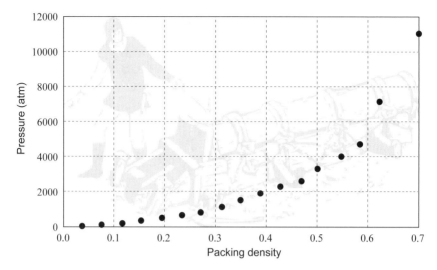

Figure 2.2. Experiments conducted by the unscrupulous but talented eighteenth-century American scientist, spy, and adventurer Benjamin Thompson (Count Rumford) showed that the amount of gun barrel pressure generated by *deflagrating* (burning) black powder depends upon how tightly the powder is packed. Packing density is 1.00 when the powder is packed as tightly as possible (too tight for use in weapons); it is 0.00 when loose. The pressures are huge; for comparison, water pressure at a depth of one mile beneath the ocean surface is 155 atmospheres. Data from Rumford (1876, 2:148).

black powder and its burn rate had a significant influence on the evolution of both small arms weapons and artillery, as we will see.

Black powder can be ignited by flame, spark, friction, or shock, although in comparison with other explosives it is remarkably stable, as Richard Whiting illustrated in a paper on its properties (Whiting 1971): "There can be no question that in the absence of moisture black powder is extremely stable in normal conditions. I am informed that until World War I it was the practice of the French Army to preserve lots of black powder which had proved especially good, for use in time train fuzes and it was reported that some lots so preserved dated from Napoleonic times. During the Williamsburg reconstruction beginning in 1926, unexploded Civil War shells loaded with black powder were unearthed from time to time, and often were found with the black powder in them still in good condition."

Ignition of propellant is a problem because, of course, the propellant is contained within a confined space. The easiest way for a gunner to ignite powder in his cannon was by introducing a source of heat—but how? Flame was found to be the most convenient method, although one of the many hazards of black powder use was the variable speed of ignition. A flame spread from one grain to the next much faster than a grain burned, so that an entire charge of loose powder tended to go off at once. On the other hand, tightly packing the charge would slow down ignition rate. Recall that slower ignition generally improved muzzle speed; tight packing changed the game by changing the pressure generated by combustion. Toward the end of the era of black powder munitions, gunners learned an empirical rule that showed how the pressure generated by a charge changed with how tightly the charge was packed (see fig. 2.2). So you can see that learning to work with black powder propellant was a long, slow process because of the number of factors that determined how the powder would burn. This learning process took centuries.[7]

Over the period of black powder propellant use, roughly the half millennium from 1350 to 1850, artillery gunners built up a large body of learning. I can best convey their progress by providing a brief snapshot of the state of artillery at both ends of this period. (In the next section I will fill in the gaps with a summary of technical developments covering artillery and small arms development over this period.) At the beginning of the black powder era, cannon were vase-shaped containers firing crude stone balls, with powdered charge of inconsistent weight and quality. Some early cannon were formed from iron bars made into a cylinder and crudely held together by iron hoops; casting long iron cannon in one piece followed later. In both cases the ball—of variable weight and shape—was dropped into the muzzle by hand. The crudeness of the cannon and the projectiles meant that *windage*—the gap between projectile and bore—was both erratic and substantial, resulting in the loss of much propellant energy.

In sharp contrast, toward the end of the black powder era well-engineered and specialized artillery pieces were provided with their own size of grain and weight of charge, with rounds of ammunition manufactured for each specific type of gun and target and loaded with the aid of numerous specialized implements. Table 2.1 provides a mid-nineteenth-century list

7. For the evolution and deployment of early firearms, see Guthrie (2002), Hall (1997), and Pauly (2004).

Table 2.1. Black powder grain size for different types of guns

Weapon	Grain size (in.)
Musket	0.035–0.06
Mortar	0.07–0.10
Cannon	0.27–0.31

Source: Data from an 1855 U.S. Army training manual.

of required grain size for different weapons. How and why did gunpowder weapons develop from their crude beginnings to hi-tech specialist artillery and small arms? Why should different weapons require different grain sizes?

THE EVOLUTION OF BLACK POWDER WEAPONS

Before we begin to examine how black powder weapons evolved in their 500-year history, some basic definitions. A firearm is a portable gunpowder weapon and includes both ordnance (big cannons) and small arms. The precise definitions get technical (for example, a cannon is a gun but a howitzer is not), so in this book I shall simplify and use the words *gun* and *firearm* interchangeably. Also, here, small arms are hand-held firearms, whereas artillery is everything else, including mortars. This simplification could cause problems with, for example, heavy machine guns or bazookas, but I won't be discussing them much and so there should be no ambiguities.

Artillery

Given that black powder weapons originated in Asia, it is not surprising that Europeans were initially behind in their use and development. In the Middle Ages, the Chinese, Indians, Persians, Turks, and Arabs all possessed more, and more sophisticated, guns than did the Europeans. The first European black powder weapon—the vase-shaped gun I referred to above—fired metal arrows. It was known as a *pot-de-fer*, and it must have been next to useless. A similar weapon, which fired stone balls in high, looping trajectories, was the *bombard*. (The Mons Meg gun on display in Edinburgh Castle, Scotland, is a large bombard dating from 1450.) The powder was ignited with a heated wire inserted into the touch-hole. The bombards had short barrels (like later mortars) so that they could be loaded from the muzzle end, even when elevated; but these short barrels, widening near the mouth, must have made them very inefficient and inaccurate.

That the bombards' short barrels worked at all was a consequence of the fine powder of the early days. We have seen that such powder burned quickly and so released all of its gas while the projectile was still inside the barrel. Because quick-burning powder generated higher pressures inside the barrel, bombard barrels (and the barrels of early artillery in general) were thick as well as short, to withstand the great stress imposed upon them by fine black powder. The later (corned) powder burned more slowly, giving lower pressure peaks; consequently, the barrels could be thinner. On the other hand, slower burn meant that the barrels had to be longer; otherwise, the projectile might have left the barrel before all the powder had burned, thus wasting powder and reducing range. Here is the first suggestion that barrel length is related to the power and range of a gun. Now you see why the mortar of table 2.1 required smaller grain size than the long-barreled cannon.

For two centuries after the introduction of black powder into Europe (say from 1325 to 1525), a great deal of experimentation took place as artillerymen learned how best to use this new-fangled invention. Cannon shapes, lengths, charges, and projectiles were all tried in different combinations, generating a confusing proliferation of names: *perriere, falconet, culverin, ribaudequin*. There were breech loaders and muzzle loaders, single-barreled and multi-barreled, vase-shaped and cylindrical, short or long; they fired arrows, stone balls, iron balls. At the end of this initial period of innovation, a small subset of these designs remained, and these would carry through to the eighteenth century, with relatively few changes. When the dust settled, cannon were muzzle loading with bronze barrels cast in one piece, firing cast iron cannonballs.[8] They were not mobile: gun carriages were developments of the sixteenth and seventeenth centuries. The cannonballs were propelled by about half to three-quarters their weight of

---

8. Breech loading—loading from the rear part of the barrel—was a good idea but ahead of its time. The technology of making strong breeches that could withstand the high gas pressures without leaking would have to wait until the end of the nineteenth century. Multibarreled guns would similarly reappear at this time.

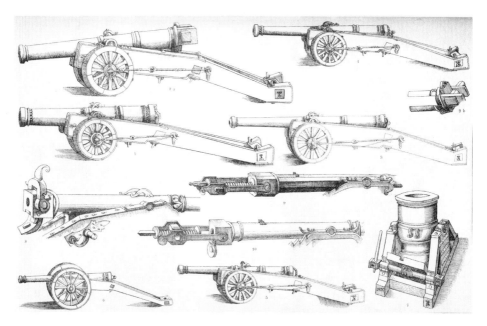

Figure 2.3. Contemporary illustration of sixteenth-century artillery pieces, including culverins, a falconet, and a mortar. Image from Wikipedia.

powder, corned to a size that depended upon barrel length and gun caliber. Figure 2.3 illustrates something of the variety of artillery weapons in the early days.[9]

These early guns were engineered to tolerances that were crude compared with later standards, and so they necessarily had substantial windage. For muzzle-loading weapons it was essential to have the ball diameter less than the gun caliber so that the ball could be loaded. On the other hand, too large a difference—i.e., greater windage—meant that much powder energy would be wasted because the gas could escape around the sides of the ball without accelerating the ball. There is a trade-off here. Technical note 5 examines this trade-off for the early firearms, where windage had to be large. In later centuries muzzle loaders and their projectiles were made to more exacting engineering standards, allowing windage to be reduced. As a consequence, the guns were more efficient.

9. See Greener (2002), Guthrie (2002), Hall (1997), McCormick (1933), and Smith and DeVries (2005) for details of early artillery development—about which, perhaps surprisingly given its early date, we know quite a lot.

It was only at the beginning of the sixteenth century that firearms began to be used widely for military purposes.[10] The new military guns changed the nature of fortifications because large siege cannon were capable of reducing medieval castles to rubble (ending the need for trebuchets). As a result, forts became more angular to deflect shot (see fig. 2.4) and lower to present a smaller target. Walls became thicker; breastworks and redoubts replaced ramparts and towers. Also in this century cannon became more mobile. From ungainly heavyweights that were good only for static siege warfare, cannons were given wheels and trails, which enabled them to be hooked up to gun carriages; *limbers*, which enabled the gun carriages to be hooked up to horses; and *caissons*, which allowed the horses to transport cannonballs as well as cannon. A sixteenth-century German development, the *mortar*, could lob munitions over castle walls. Early attempts to make hollow projectiles filled with powder proved dangerous to the mortar crew (the projectile would sometimes explode while still in the barrel) until French bombardiers solved the problem in the first half of the seventeenth century. The *howitzer* was developed as a cannon-mortar hybrid. It sent a heavy projectile in a high trajectory, though not as high as a mortar trajectory. It had a short barrel (a Dutch innovation) so that it could be loaded by hand when used at high-elevation angles. Its range was between that of cannon and mortar.

So, a couple of centuries after black powder was introduced into Europe, guns had been developed and specialized into two basic types: artillery and small arms. Artillery had split into three varieties: cannons, howitzers, and mortars. The cannons were long barreled and heavy, to take a big charge so that they could fire solid shot a great distance. Howitzers were high-caliber, high-trajectory, lighter-weight, and lighter-charge variants that fired hollow shot. Mortars were shorter, lighter weapons that lobbed hollow shot at very high elevations. The growth in numbers of artillery weapons demanded considerable organization in European

10. Even so, the French Renaissance writer Montaigne felt the need to say, as late as 1585, that "the effect of firearms, apart from the shock caused by the report, to which one does not easily get accustomed," was so insignificant that he hoped they would be discarded (Greener, 2002). Cannon of the fourteenth and fifteenth centuries were still experimental, and powder was expensive, at least until the late 1400s, as we have seen. During this period, cannon competed with trebuchets, and hand-held black powder weapons competed with crossbows.

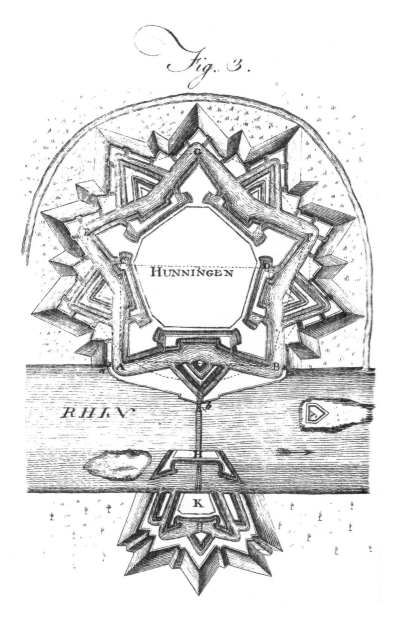

Figure 2.4. Plan of a fort at Hüningen, on the Rhine, built during the reign of Louis XIV of France by the great French military engineer Vauban. During the seventeenth century fortifications were being rebuilt to withstand bombardment by large siege cannons. Note the angular "star" design, intended to deflect enemy shot and minimize the amount of *dead ground*, or blind spots unreachable by defensive guns. Image from *Encyclopaedia Britannica*, 1st ed., vol. 2, plate 85.

Figure 2.5. Naval 36-pounder long cannon in the Age of Sail (late eighteenth and early nineteenth centuries). From a drawing by Antoine Morel-Fatio (1810–71).

armies; logistical skills developed along with the guns, so that the many and various artillery pieces, each with the correct carriages, ammunition, and powder, could be delivered to the right place in a timely manner.

The seventeenth century brought in cannon that could be *aimed*. Previously, artillery could hit nothing smaller than a large castle, but by now the range of cannon had increased enough to require aiming, and technology had improved sufficiently for it to be practicable.

In the eighteenth century the *carronade* was introduced. The name may sound like a cross between a Christmas carol and a serenade, but the carronade was anything but a lullaby. It was a short cannon with low muzzle speed that was designed for the British Royal Navy. We will see in a later chapter how effectively cannonballs demolished wooden ships. Suffice it to say here that the slower-moving (and heavier) carronade balls caused wood to splinter more, thus spreading death and destruction among the crew of enemy ships. The relatively small charge of these guns gave them a much shorter range than the long-barreled cannons (fig. 2.5), but it also meant that they were less expensive and much less heavy—a significant factor in Age of Sail naval warfare.

As artillery design and engineering improved, and as understanding (albeit empirical) of gunpowder grew, less and less weight of charge was required to fire a projectile of a given weight to a given range. In the earliest days the weight of powder was typically about the same as the weight of the projectile. By the early sixteenth century it was about two-thirds of projectile weight and reduced to a half by the end of the century. It gradually reduced further—down to a third in the eighteenth century—until, in the mid-nineteenth century, the charge was between a quarter and a sixth of projectile weight.

Nineteenth-century artillery developments were relentlessly in the direction of bigger guns firing heavier and more destructive projectiles to longer ranges with greater accuracy. I will defer further discussion of late-nineteenth and twentieth-century cannons, and modern artillery, until the next chapter.

Small Arms

Now we backtrack to the fifteenth century and follow the evolution of the second main branch of black powder weapons. Small arms, ranging from the hand-held culverin to the *arquebus*, were developed. The arquebus had a stock and was a recognizable antecedent of the modern rifle. A smooth-bore, muzzle-loaded matchlock weapon, it dominated hand-held gunpowder weapons from the fifteenth to seventeenth centuries.[11]

The arquebus was succeeded by the larger musket, a weapon that is familiar to many people because it lasted so long (from the seventeenth to the nineteenth centuries) and because it was so widespread. It began as a long smoothbore, muzzle-loaded wheel-lock weapon that was heavy enough (it was heavier than the arquebus, and longer, during the period when both were in use) to require a bipod to support the barrel, and ended as a rifled muzzle loader that was fired with a percussion cap and fixed with a bayonet. The original was accurate out to perhaps 60 yards (though useful up to 200 yards), whereas the late-model muskets were accurate to 200 yards and good out to 500 yards. This later (rifled) musket would be succeeded by the breech-loading rifle.

Small arms used finely grained powder because of their short barrels, compared with artillery, but also because of their small caliber. They needed a much smaller charge to accelerate a lead ball to high speed, and

11. For the development and influence of early firearms, see Hacker (2006, chaps. 1–2), Hall (1997), and Pauly (2004).

so the pressures generated by charge ignition were much lower than those of cannon, even for quick-burning fine powder. Of course, firearm barrels were much thinner than those of artillery but were easily strong enough to withstand the pressure of fast-burning powder. This is so because of the effects of scale, explained in technical note 6.

Ignition of a black powder charge was a particular problem for hand-held weapons. It had been solved for artillery by introducing a touch-hole to which a gunner would introduce a flame when the gun was to be fired. The flame would ignite *primer* powder, which would then—via the touch-hole—set off the main charge.[12] The artillery piece had already been aimed, of course, by gun-layers. For a small arm, however, there was only person to do everything, so how could he aim his arquebus or musket and at the same time ignite the primer?

The first solution, which lasted for a couple of centuries, was the *match-lock* system. A slow match, consisting of a length of thick string that had been soaked in saltpeter and lit at one end, was transferred from a holder (the *serpentine*—so called because it was S-shaped) on one side of the barrel to the touch-hole via a triggerlike lever that was borrowed from the crossbow. Pulling the lever caused the lit end of the match to fall onto the primer pan. The musketeer was required to blow on the slow match (to cause it to glow) prior to aiming. Needless to say, the system was unreliable in windy or wet weather. Loading a musket in those days was a complex operation that required much training (the origin of military drill).[13]

Succeeding the matchlock was the *wheel-lock* system (though they overlapped for some time). Here the slow match was replaced by a spring-loaded serrated steel wheel that rubbed against a block of iron pyrite when released by the trigger. This rubbing caused sparks that would ignite the primer. The primer charge was held in a flash pan—a small open container (normally covered to protect the primer from the weather) that was uncovered by the action of the trigger. (Our expression "flash in the pan" comes from this device.) Wheel-lock pistols (a sixteenth-century example is shown in fig. 2.6) were the first firearms that could be carried by light cavalry, and this combination—pistol and horseman—was the death knell

12. Primer powder was loose and fine-grained, so that it burned easily. The flame would carry down the touch-hole to ignite the main charge.

13. Modern musketry trainees can consult YouTube videos. For example, see three videos cited in the bibliography (YouTube 2007a, 2007d, 2008a) for demonstrations of percussion cap, matchlock, and flintlock musket operation.

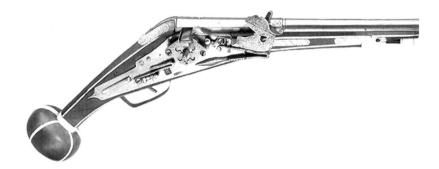

Figure 2.6. A wheel-lock pistol of 1580. Adapted from a photo by Nick Michael.

for medieval heavy cavalry. The effective range of these pistols was only a few yards, and they could not penetrate the thickest armor, but they meant that poorly armed horsemen could fell expensively equipped knights.[14]

The wheel-lock was quickly followed by the *flintlock* system, shown in figure 2.7. Invented in the late seventeenth century, probably in Spain, flintlock muskets lasted until the introduction of percussion caps in the late 1800s. Flint replaced iron pyrite. The flint was held in a spring-loaded arm called a "cock," which was pulled half back to enable the primer to be placed in the pan (giving rise to the expression "going off half cock"). The cock was then pulled back fully prior to the trigger releasing it and causing the flint to spark, thus igniting the primer. This system was simpler and more robust than the wheel-lock.

The *caplock* system had to wait for developments in chemistry. The explosive salts known as fulminates were developed in the nineteenth century. These pressure-sensitive explosives can be readily detonated by percussion. The first of these used in musket ignition was mercury fulminate. In the caplock system a cock or hammer replaced the flintlock and priming pan. This spring-loaded cock would be released by the trigger and strike a thin metal cap containing fulminate, which would then detonate, sending hot gas down the touch-hole to ignite the charge. It was a simpler method than any previous system, requiring less training for the muske-

14. Firearms were a great leveler, in the sense that any soldier could kill any other. By the time the wheel-lock system was available (the sixteenth century), pistols were less expensive than longbows or crossbows and required less training. No longer were knights—lavishly equipped with armor, a horse, and a lifetime of training—safe from the ill-trained common soldier.

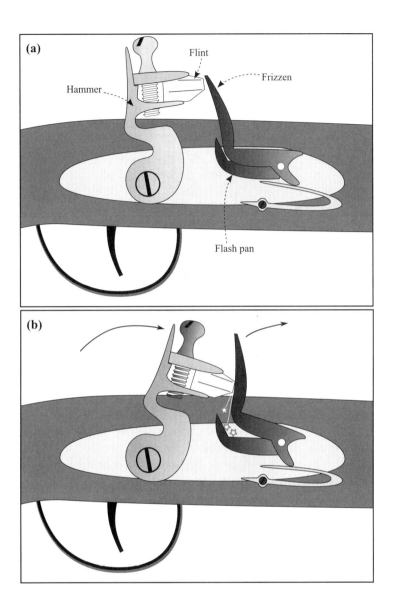

Figure 2.7. How a flintlock works. (a) The hammer is shown in the half-cock position. The frizzen (a serrated metal projection) is pulled back to reveal the flash pan, which is filled with primer before the frizzen is returned to the position shown. Then powder and ball are rammed down the muzzle. The hammer is pulled back to full-cock; the musket is aimed. (b) Pulling the trigger sends the hammer forward, where the flint stikes the frizzen, sending sparks onto the now-uncovered flash pan and igniting the primer. A hole connects the flash pan to the charge, which ignites and sends the musket ball on its way.

teer and permitting more rapid firing. It was also more robust and was less susceptible to wet weather. By this time the use of *cartridges* was widespread. (They had been invented in 1744, perhaps by the great Swedish military leader, King Gustavus Adolphus.) A cartridge in those days consisted of a paper wad containing a musket ball and a black powder charge. This simplified the loading process and ensured that the correct amount of charge was employed.[15]

So, by the time of the U.S. Civil War the three components of musket operation—ball, charge, and primer—had been reduced to two physical components (cartridge and percussion cap). The next step unified all three into a single entity: a modern cartridge or round. The consequences of this new development take us into the late nineteenth century; they are far-reaching and form the subject of chapter 3.

THE MILITARY REVOLUTION

Viewed over a period of several hundred years there is no doubt that black powder radically changed the course of human history. These changes are often acknowledged by referring to the "gunpowder revolution." While understandable as a statement of the importance of black powder to military history, this phrase is unfortunate in that it suggests the impact of black powder weapons was sudden. This was not the case. There was a "military revolution" that came about in part because of black powder weapons, but this occurred two and a half centuries after black powder was introduced into Europe, and the contribution of black powder to the revolution was patchy. It is argued that pre-gunpowder weapons were limited in design by the strength of the weapon user, whereas the design of gunpowder weapons was free of such considerations so that firearms could be designed according to tactical needs. This is why, the argument goes, firearms generated the military revolution. Viewed long term, this statement is true, but a participant of the revolution could be forgiven for not even noticing that it was happening. Popular understanding of history often overlooks the odd stop-start contribution of firearms to the revolution.

First, some bare facts about the military revolution. Over a period of a few decades in the second half of the sixteenth century, warfare in Europe changed significantly. Armies grew in size and became more professional. Battle casualties increased (i.e., the number of casualties, as a fraction of

15. See YouTube 2007b for a video showing caplock operation.

troops involved, increased). Warfare had more of an impact on societies. These are the facts; what is in doubt is the degree to which these changes were brought about by firearms. We have seen that, during their first two centuries, firearms evolved in many directions as weapons designers sought the best ways of exploiting the new black powder. However, the military revolution rook place at a time when firearms evolution was slowing down, after the development of corned black powder and of the wheel-lock pistol. So what role did firearms play?

The first major impact of black powder weapons was in siege warfare. Early cannon made short work of medieval castles, which were designed to resist trebuchet stones, not the relatively high-speed stone balls fired from large guns. Sieges became shorter—a significant development. Then new fortifications, designed especially to withstand cannon fire, were built to replace castles. This restored the status quo; sieges became as time-consuming as they had been formerly. So cannon were not responsible for the military revolution.[16]

Increased casualty rates were probably due to firearms. The damage done by early ballistic weapons was not selective or controllable. A bombardier or musketeer could not choose to merely wound, not kill, an enemy and then take him prisoner for ransom in the medieval fashion— he pointed his weapon in the general direction of the enemy and fired. The large increase in casualties was particularly severe among the rich and influential leaders, a fact that was commented upon at the time. These people would formerly have been disarmed and ransomed, but a cannonball does not stop to inquire about disposable income.

The increasing size of armies was in part due to nonmilitary reasons. At the beginning of our period the general population was rebounding from the effects of the Great Plague. Later, economic hard times drew peasantry into military service in large numbers.[17] On the military side, the growth of armies was also an indirect consequence of the effectiveness of pikemen

16. Cannon were not very mobile in the first three centuries of their development, so their main application was to siege warfare.

17. The motivation for peasants to join armies in the late medieval period has been compared with that of people who buy lottery tickets today (buyers are predominantly working class). Though their lives are unlikely to be improved as a consequence, well-publicized stories of people who become very rich in this manner encourage more people to buy lottery tickets. See Hall (1997) for this analogy and for much thoughtful commentary about firearms of this period. Parker (2000, chap. 6) provides a more wide-ranging account of warfare development.

against cavalry; dense pike formations were introduced to many European armies during this period. The effectiveness of this tactic, combined with the immobility of gunpowder weapons, led to defensive warfare. Defenders would usually win a battle, so battles were avoided. Consequently campaigns were decided by territorial occupation, which required the service of lots of soldiers. Here we find reasons for the greater impact of warfare on society: the tendency toward occupation was felt by the wider population. The increased tax burden and the logistics of supplying larger armies was felt by all.

One of the few clear-cut examples of firearms' influencing the military revolution is supplied by wheel-lock pistols. When these pistols became inexpensive enough to be widely distributed, they led, as we have seen, to the demise of heavy cavalry. This demise led to the disbandment of pike formations, which in turn had consequences for the dominance of defense.

THE DEVELOPMENT OF UNDERSTANDING

A lot of the improvements in black powder weapons that I have summarized in the preceding sections were motivated by military necessity. The consequences for a country of losing a war were disastrous for national prestige and well-being. In an age when European wars were frequent,[18] there was great incentive for improving military technology. Much of this improvement was empirical—by trial and error—but in the eighteenth century the understanding of military technology began to be placed upon a more scientific footing. (Trial and error can be a perfectly valid scientific approach: it combines experimental variation with intelligent interpretation of results.)

In 1742 a British mathematician and military engineer named Benjamin Robins published a book called *New Principles of Gunnery*. In it, he showed the results of experiments that he had conducted on numerous firearms to establish, for example, the relationships between gun caliber, barrel length, powder charge, and projectile muzzle speed. These measurements were made with a *ballistic pendulum*, a device that he invented to estimate muzzle speed. The use of ballistic pendulums spread far and

---

18. For example, the eighteenth century consisted of a long-running series of wars between Britain and France—fought on three continents and all the world's oceans—for dominant position. Final score: Britain 4, France 1. (The single French victory was in the American Revolution.)

wide, and the original design lasted for more than a century before being superseded by an electronic measuring device, the *chronograph*, which operated on different principles. The ballistic pendulum, and Robins' scientific approach, have led to his being widely acclaimed as the father of modern ballistics. Robins by no means solved all of the many and varied problems of internal ballistics, but he showed us how to proceed. Leonhard Euler, the famous Swiss mathematician of the late eighteenth century, critiqued and expanded upon Robins' work. His mathematical approach did much to transform ballistics into the mathematical science that it is today. Between them, the two men turned a trial-and-error field into a scientific discipline.

There were also earlier scientific investigators. These innovators included Charles V, sixteenth-century Holy Roman Emperor, who tested the range of a long culverin. The original length was 58 calibers, but this was successively decreased to 50, then 44, then 43 calibers, and it was found that the range increased with each reduction. At the time, this finding must have caused some confusion because the general belief in those early days was that the range of an artillery piece increased with increasing barrel length (assuming that the same weight and quality of charge was used in all cases). Consequently, many culverins were of enormous length. One reason for the belief was that it seemed to be true for small arms that range or muzzle speed increased with barrel length.

In the nineteenth century more precise experiments showed that there was an optimum bore length, in calibers; for barrels longer or shorter than this optimum the muzzle speed (and therefore the range) was reduced. Further, this length depended upon the charge. Also, the optimum length for a barrel (measured in calibers) was found to be greater for small arms, which fired lead shot, than for cannon, which fired solid iron balls. The optimum length for cannon was greater than the optimum lengths for howitzers and mortars, which fired hollow shot. So it seemed that optimum barrel length increased with projectile weight or density and depended upon the power of the charge.

Capt. J. G. Benton, of the U.S. Army, writing in the mid-nineteenth century, quoted results of experiments carried out by one Major Mordecai with a 12-pounder cannon. Mordecai found that muzzle speed increased with barrel length up to a length of 25 calibers, but that the increase beyond 16 calibers was very slight (about 5%). Benton went on to graph the results of later tests with a similar artillery piece (reproduced here in fig. 2.8) that showed how muzzle speed depended upon barrel length and

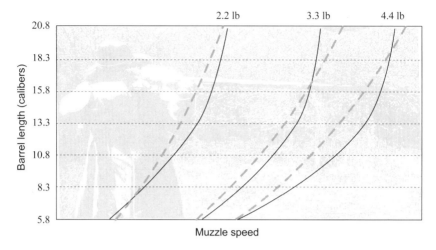

Figure 2.8. Gun barrel length vs. muzzle speed for nineteenth-century U.S. smoothbore cannon. The black lines are results from Benton (1862). The mathematical model of technical note 7 predicts the curves shown as dashed gray lines.

how the plotted curve changed with the weight of charge. These results he summarized empirically by saying that muzzle speed, v, increased as the quarter power of barrel length, L, or $v \sim L^{1/4}$, roughly.

Can we make sense of all this data with a simple physics explanation? Indeed we can. Technical note 7 shows how the internal ballistics of black powder firearms—either small arms or artillery—explain (approximately) all of these findings. This technical note gets to the heart of internal ballistics. In the spirit of the approach discussed in the introduction, the calculations involve simplifying approximations rather than exact but complex equations because the simpler approach is more revealing. I am aiming for clarity, but the full plethora of complicating real-world details can obscure, much as black powder smoke obscured a battlefield.

The calculation of technical note 7 can be applied to a typical mid-nineteenth-century firearm for which we have data: the 1855 Springfield rifle-musket (see table 2.2). We estimate that the time interval t_m between charge ignition and the bullet's emergence from the end of the barrel is about 5 ms (five-thousandths of a second), which is close enough to measured values to vindicate the calculation of technical note 7. (I expect only ballpark accuracy from my calculation, if not from the Springfield.) Table 2.2 shows some of the parameters applying to other nineteenth-

Table 2.2. Data for some nineteenth-century American small arms

Weapon[a]	Gunpowder mass, m_g (gr)	Projectile mass, m (gr)	Muzzle speed, v (ft/s) Real	Estimated[b]
Springfield	65	500	950	
Carbine	55	450	820	925
Colt	14	125	760	885
James	100	217	1,900	1,800

Source: Data from an 1855 U.S. Army training manual.
[a] A carbine is a shortened rifle, the Colt is a pistol, and the James is a sporting rifle.
[b] Calculated from the model of technical note 7. The estimates are pretty good, considering the simplicity of the calculation.

century small arms, which enables us to further apply the predictions of the simple math model constructed in technical note 7. From the Springfield data we fix the unknown constants kf of equation (N7.10) as $kf = 215,000$ m^2/s^2 and apply this value to the other weapons;[19] the predicted muzzle speed is compared with real data in the table. The worst estimate is in error by 16%. So the simple approach of technical note 7 gives us reasonable numbers.

Finally, we can adapt the analysis of technical note 7 to explain a couple of odd facts that emerged from the history of ballistics investigations. Nineteenth-century French tests on a French 36-pounder cannon showed that increasing the charge beyond the optimum amount caused a *drop* in muzzle speed. This seems counterintuitive at first, but a little thought shows why. Not all the powder would have burned, and the unburned powder would have been expelled from the cannon (as is discussed in technical note 7). So the powder that did burn caused both the cannonball

19. In applying the kf value obtained for one firearm to all the firearms in table 2.2, I am making assumptions. The fraction, f, of powder mass that is converted into gas can safely be assumed to be about 50% for all these weapons, so no problems there, but the value of k may vary. This parameter, introduced and discussed in technical note 7, depends upon packing density and temperature—which may vary from weapon to weapon. The powder mass (as a fraction of projectile weight) used for the firearms of table 2.2 is less than for artillery of the period: mid-nineteenth-century firearms were better developed than artillery at that time, as we will see in the next chapter.

and the unburned powder to accelerate. Accelerating unburned powder is a waste of energy. (Recall that the mass of the powder was a significant fraction of projectile mass.) Also, a large charge would take up more space (l of fig. N6 would increase), thus decreasing the effective length of the barrel, from cannonball to mouth. With reduced L, the muzzle speed reduces. Turning to Charles V of Spain, three centuries earlier: we left him puzzling why his longer barrels caused reduced muzzle speed. The answer is friction, as explained in technical note 8.

Stepping back a little from the detailed calculations of technical notes 7 and 8, we can see the big picture of internal ballistics by thinking of the charge in a gun's chamber, and the projectile in the gun barrel, as a piston about to begin its power stroke. The sudden expansion of gas from the burning powder causes the projectile to accelerate along the barrel, just as the ignition of fuel causes a piston to slide along its cylinder.[20] What determines the power of a piston is the swept volume, that is to say, the volume of cylinder that the piston head moves through during its cycle. The same is true for the projectile of a gun: it obtains energy from the propellant all the while it is inside the barrel. So, large projectile energy requires large swept volume—in other words, high caliber and/or a long barrel. (This is why short-barreled handguns tend to have higher calibers than rifles.) But there is a limiting factor. A piston rod limits the length of the piston stroke, and friction limits the useful length of a barrel. So long as propellant gas pressure pushes the projectile harder than friction holds it back, energy is being added to the projectile; otherwise, energy is being subtracted. Thus, for a given charge there is an optimum length for a gun barrel, as we found in technical note 8.

Black powder is a mixture that *deflagrates* upon ignition; the best form for this mixture depends upon the gun and was found by trial and error over centuries. Similarly, the best design for guns was found by empirical tinkering—trial and error—particularly during the first two centuries these weapons were in use. A simple mathematical model of internal ballistics can explain many of the features observed in real firearms, such as the dependence of muzzle speed upon barrel length.

20. Interestingly, in the seventeenth century a very talented Dutch physicist, Christiaan Huygens (a contemporary and rival of Sir Isaac Newton), considered using gunpowder to drive a piston, thus creating an internal combustion engine. His ideas were way ahead of the technology of his time, however, and the Huygens engine could not be built.

3 The Development of Modern Firearms
New Technical Challenges

We begin this chapter where we finished chapter 2, with rifled muskets in the middle of the nineteenth century, and will progress from these strange hybrid shoulder-fired weapons to their modern equivalent: assault rifles. In parallel, we will look at the development of ordnance over the same period. The reason for pausing at this particular time in history (around 1850) is because the succeeding half century saw the greatest changes and the most rapid advances in firearms technology—more rapid than any period before or since. You can readily appreciate this surge in weapons construction and capability (which ultimately was due to the Industrial Revolution) by considering the state of firearms at the beginning of the nineteenth century, and the beginning of the twentieth. In 1800 small arms and ordnance consisted mostly of smoothbore muzzle loaders firing spherical ball projectiles propelled by black powder. By 1900 we had automatic weapons firing metal-jacketed cartridges propelled by smokeless powder.

Since then, developments have been significant but more incremental, less revolutionary. Metallurgy and propellant chemistry have led to improvements in the muzzle velocities and ranges of firearms. Twentieth-century refinements in design have led to increased calibers and rates of fire—important improvements, but nothing like the unprecedented leap forward that occurred during the five decades between 1850 and 1900. The twenty-first century may well produce another surge in firearm development, though one that is less closely related to our subject of ballistics: computer processing and remote sensing technology will revolutionize (are already revolutionizing) the long-range accuracy of missiles.

The internal ballistics of firearms changed apace with the technological advances. We can see why, and how, by going back to the mid-nineteenth century and tracing the main developments.

HISTORY IN THE FAST LANE:
THE SMALL ARMS REVOLUTION

In 1853 the Enfield rifled musket was introduced in Britain; two years later the Springfield rifled musket appeared in America. Both these weapons would be produced in massive numbers, and both would evolve over the course of their distinguished careers. They existed because in 1849 Claude Minié in France made the crucial breakthrough in bullet design that spelled the end for smoothbore muskets. Yet rifled muskets were stopgap, transitional weapons. The writing on the wall, pointing to the future, occurred at exactly the same time. In 1853 Christian Sharps, in America, produced his famous breech-loading rifle (actually a percussion carbine).[1]

The Enfield rifled musket was used by the British during the Crimean War (1854–56), where it is said to have given British soldiers a great advantage over their Russian opponents, and again in the massive Indian Mutiny of 1857. The Springfield was the standard weapon of the American Civil War: over 1½ million were used (and half a million imported Enfields; they were almost the same caliber and so could use the same ammunition). Rifled muskets had a slower rate of fire than smoothbore muskets but were much more accurate and were effective out to a much greater range (over 250 yards in the case of an 1855 Springfield, compared with about 75 yards against an individual target for a late-model smoothbore musket). There are two reasons for the difference in performance: rifled barrels and Minié bullets.[2]

Rifling refers to the helical grooves cut along the inside surface of the barrel. (Some technical aspects of rifling are discussed in technical note 9.) Known in Europe from the sixteenth century but little used for 400 years, for reasons that will soon become clear, rifled small arms were probably developed as a response to the fouling of barrels that resulted from burning black powder. Recall that half the chemical products of black powder deflagration are solid; some of this solid material would, after a

---

1. Pauly (2004) is good on small arms development during this period. See also Berger (1979).

2. Benton (1862) provides an interesting account of the state of firearms in the mid-nineteenth century, including details of the 1855 Springfield rifled musket. For historical development in small arms, see Britannica (1998, s.vv. "Bullet," "Cartridge," "Rifled Muzzle-Loaders"), Pauly (2004), and Hacker (2006).

number of firings, gum up the barrel. Rifling provided a place for the gunk to accumulate, so that muzzle loading and projectile speed would not be compromised too badly. That was the theory. Why do we think this is the original motivation for rifled gun barrels? Several early small arm weapons have been found with *straight* rifling—no twist. It was soon discovered that twisted rifling made the firearm more accurate; we now know that this is due to gyroscopic stabilization (which I will discuss in chapter 5). Beginning in the eighteenth century, some rifled muskets were produced and used in warfare,[3] but there were always fewer of these than of smoothbore muskets until the 1850s.

To work properly, so that it develops a spin as it accelerates down the barrel, a musket ball must grab the rifling grooves—and so the ball must fit snugly. This made muzzle-loading a pain because the ball had to be rammed down the barrel. (Ramming also tended to compromise the improved accuracy somewhat by distorting the soft lead ball, making it fly erratically.) So, as I have already noted, the rate of fire of rifled muskets was less than that of smoothbore muskets even though their effective range was greater because of the tight seal—no windage. On an eighteenth-century battlefield with ranks of closely packed infantry, the inaccuracy of smoothbore guns was not so important; similarly, the vast amount of smoke generated by musket volleys meant that the rifled musket's longer range was not much of an advantage. Thus, in the eighteenth and early nineteenth centuries rifled muskets were relegated to use by hunters and skirmishers.[4] The main business of battle was conducted with smoothbore weapons.

All that changed with the Minié bullet. The end result of a series of developments from about 1825, mostly in France, the Minié bullet was shaped like a modern bullet (called at the time a "cylindro-conoidal ball") not like the spherical musket ball. The bullet's diameter was less than the caliber of the rifled musket for which it was designed, and yet there was no loss of

3. The British had several rifle regiments in the Peninsula War (part of the Napoleonic Wars that took place in Spain and Portugal). The Russians deployed as many as 20,000 rifles against the French, when the latter invaded in 1812. See Pauly (2004) for the early use of rifles and for the serendipitous discovery of the benefits of rifling.

4. In America the celebrated Pennsylvania and Kentucky rifles formed part of this tradition of early rifled firearms. They were famously accurate at long range; they also followed the general trend of reduced caliber, compared with contemporary smooth-bore muskets.

power due to windage. This was because the back end of the Minié bullet was hollowed out, so that the gas pressure produced by deflagrating powder would cause the end of the bullet to expand, grabbing the rifling and sealing the barrel. This new bullet design solved all the problems of muzzle-loaded rifled small arms: easy loading because the bullet diameter was less than the caliber and no compromising of accuracy from distorted bullet shape because the bullet did not need to be rammed into the barrel. Furthermore, the bullet shape is more aerodynamic than that of a musket ball; for this reason, as well as the elimination of windage, rifled musket ranges improved.

Add to the improved range and accuracy an improved ignition system—caplock instead of flintlock—and you have the difference between a Napoleonic musket such as the Brown Bess and a Civil War rifled musket such as the 1855 Springfield.[5] The increased range and accuracy of these small arms changed warfare. Thus, to give one small example, colorful uniforms disappeared, to be replaced by cryptically colored gear that rendered a distant soldier less visible.[6]

Artillery development lagged behind small arms development by several decades, for reasons we will soon see, and so the American Civil War was dominated by the rifled musket. This new weapon changed the tactics that were required for victory on the field of battle: no longer could massed ranks of infantry advance in formation over open ground. Civil War commanders were slow to realize that the new technology was changing the face of war, and so casualties were high. This factor—that tactical developments lagged technological ones—would culminate in the slaughter of World War I, fought at a time when artillery (but not military thinking) had absorbed the new developments.

Just a few years separated the American Civil War from the Franco-Prussian War (1870–71). This war was also dominated by small arms and was also very bloody. The small arms, however, were no longer rifled muskets but were instead breech-loading rifles. Better manufacturing techniques (in which quality-controlled factory production replaced older craft traditions) and new rotating bolt action breechblocks (fig. 3.1a) pro-

---

5. Caplocks replaced flintlocks very quickly, in part because it was easy to convert a flintlock rifle to caplock.

6. Hunters and skirmishers were always less gaudily clad than regular soldiers for this reason. The rifle regiments under Wellington's command in the Peninsula War were later called Greenjackets because they substituted green for the more usual British red coats.

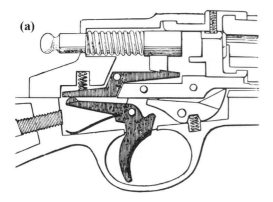

Figure 3.1. Rifle action. (a) Schematic of a bolt-action weapon. Taken from an illustration in *Encyclopédie Larousse Illustré* of 1897. (b) Winchester model 1873 lever-action short rifle. In this repeater rifle with a tube magazine in the stock, the trigger guard pulls down and forward to eject a spent cartridge case and load a new round. Photo by Bob Adams (www .adamsguns.com).

duced an effective seal that eliminated the old problem of propellant gas leaking from breeches. The Sharps rifle, appearing at the end of the U.S. Civil War, was a single-shot breech loading rifle, but the French Chassepots and the less robust Prussian "needle guns" were repeating rifles, and these cut swathes through enemy infantry. The new technology that led to repeaters took the form of metal cartridges and a spring-loaded magazine to hold them.

Earlier cartridges consisted of a paper wad containing powder and ball—the cap was separate. A soldier loading his rifled musket would bite the end off the cartridge; pour the black powder down the barrel, followed by paper and ball or Minié bullet (the paper wadding acted as a seal, reducing ball windage somewhat); place a percussion cap under the cocked hammer (later replaced by paper tape caps); and then aim and fire, achieving perhaps three shots per minute. The new metal cartridges contained powder, bullet, and primer. A lever action (see fig. 3.1b) loaded a cartridge into the chamber; releasing the trigger caused a firing pin to strike the cartridge base, thus igniting the primer, which, in turn, set off the powder. The highly successful American Spencer repeating rifle of this period, and the Henry, could fire 20 rounds a minute. Pulling the lever down and forward ejected the spent case; pulling it back into place injected a fresh cartridge into the chamber—and no need to take your eye off the target. The British Martini-Henry rifle of 1871 had a similar lever action and was accurate up to 600 yards.

Handguns also were revolutionized by metal cartridge rounds. A revolving magazine to store six of them gave rise to the revolver—the "six-shooter" beloved of Western fans. A Colt invention—the "crane" that is used to swing out an empty magazine cylinder—made these weapons quick to load. Adam and Beaumont in England developed the double action revolver; readily taken up by Colt and by Smith and Wesson, this development made revolvers quick to fire and to load. A double action mechanism provides the shooter with a choice. For rapid fire he can repeatedly pull the trigger, which cocks and fires the gun, ejects the spent cartridge and loads the next round, all from the same trigger pull. This pull was heavy, and so aiming was not easy, but if you need rapid fire at close range (say you are involved in a gunfight at the OK Corral), then this is the option you choose. For greater accuracy you pull back the hammer with your thumb to cock the weapon before pulling the trigger.

The 1880s brought another key development: after half a millennium, black powder was replaced by a new propellant that became known as *smokeless powder*. Guncotton (made by soaking cotton in concentrated nitric acid) had been invented in the 1840s but proved to be unsuitable as a propellant; however, improvements by Paul Vieille in France led in 1884 to his Poudre B, which was revolutionary.[7] When ignited, each gram of this new powder rapidly produced about a liter of gas, making it three times as powerful as the old black powder. Furthermore, because it converted almost all of its mass into gas, there were fewer solid products of burning, which meant less fouling of the gun barrel and—importantly for military purposes—much less smoke. Hence the name.

Hard on the arrival of smokeless powder, the French army introduced the Lebel rifle (1886), the first to take advantage of the new propellant.

---

7. *Poudre B* (for *Blanc*) is French for "white powder," in contrast to the traditional black powder. Poudre B is essentially gelatinized gun cotton with added ether and alcohol. It was after the introduction of smokeless powders that the name of the old propellant was changed from "gunpowder" to "black powder." Here I use "gunpowder" to refer to the smokeless powders. See Heramb and McCord (2002) for the development of smokeless powder.

This rifle was in service for three decades, and nearly three million were made. Its 8 mm bullet left the barrel at a muzzle speed of 2,300 ft/s (700 m/s) which yielded a maximum range of 4,500 yards (4.1 km). Note the small caliber: 8 mm is 0.315 inches—much less than the 0.577 inches typical of rifled muskets (which is, in turn, less than the 0.75 inch caliber of most smoothbore muskets). Stronger propellants meant less powder per charge, and so smaller calibers resulted. Two immediate consequences for the infantry soldier: his firearm was lighter, and he could carry more ammunition.

Within a couple of years of Vieille's invention came two more smokeless powders—ballistite and cordite, both developed in Great Britain. These proved safer to handle than Poudre B (which became unstable when stored for long periods). A key component of smokeless powders is nitroglycerine, an explosive that forms up to 50% of smokeless powder. (Smokeless powder containing pure nitrocellulose, derived from guncotton, is called *single-base* powder; nitrocellulose plus nitroglycerin produces *double-base* powders.) Other additives ensured that smokeless powders were stable and were readily formed into granules of desired shape. The old idea of corned black powder still applies to these new gunpowder formulations: they burn from the surface, and so grain shape and size determine burn rate, which, as we have seen, greatly influences muzzle speed. In fact, because smokeless powder was a chemical rather than a physical mixture of different particles, it was more easily molded into granules of the desired shape and size, with reliable burning characteristics.

In technical note 10, I demonstrate how different granule shapes lead to different rates of generating propellant gas. These different rates are suitable for different gun barrel lengths. As for black powder, the finer, faster burning gunpowder formulations are more suitable for short-barreled firearms such as handguns. Weapons with longer barrels benefit from slower burning gunpowder. There is a limit, of course, to how slow a powder should burn: too slow and it won't work at all; slower than optimum and the bullet leaves the barrel before all the powder is burned, reducing powder efficiency.

FIELDS OF BATTLE

Two "fields"—the Springfield and the Enfield—are the subject of this section. In particular, we will see why they changed the nature of war by looking at how they influenced tactics during the American Civil War.

Figure 3.2. An Enfield (*top*) and a Springfield 63 caplock rifle (*bottom*). Springfields were the most common firearm in the American Civil War, and Enfields were second most common. Image adapted from a photo by Mike Cumpston.

These two weapons (shown in fig. 3.2) had more similarities than differences even though they originated in different parts of the world. Their similarity is a technological example of what biologists call "convergent evolution," in which species adapt to best suit their environments. For example, sharks and porpoises resemble one another because they share the same oceanic environment, though they evolved from very different ancestors. The arms needs of the Americans and the British in the mid-nineteenth century were different, their armies and soldiers were different, and their arms production facilities were quite independent of one another—yet, when both were presented with the Minié ball and had to produce a rifle to fire it, they came up with very similar weapons.

The Springfield Armory, in Massachusetts, made over 1.5 million rifled muskets of all types—about the same as the number of Enfields made. The most common weapon of the Civil War was the Springfield Model 1861 (about 700,000), followed by the Pattern 1853 Enfield (about 500,000). As you can see from table 3.1, the two weapons were quite similar, and

Table 3.1. Comparison of Springfield and Enfield rifles

	Springfield	Enfield
Length		
Total	56 in.	55 in.
Barrel	40 in.	39 in.
Bullet size		
Caliber	.58	.577
Weight	500 gr	530 gr
Muzzle speed	950 ft/s	900 ft/s
Propellant charge	65 gr	68 gr

they were also comparable in field performance, though some officers were biased against the Enfield. Each had an effective range of 200–300 yards; this range was determined more by the limited training of Civil War soldiers than by the weapons' inherent capabilities.[8] The rate of fire in both cases was about three shots per minute, and they could fix the same bayonet. The primer systems were a little different (varieties of caplock); the Enfield's rear sight was more finely adjustable; the Springfield may have been a little more robust.

How did the rifled musket—of either type—influence the Civil War? D. H. Mahan, a key military thinker in the United States before the war, suggested that tactics needed to change to address the improved accuracy and range of rifled muskets over the older smoothbore muskets—but standard military procedures were slow to change. An army manual at the time suggested that the only necessary infantry adaptation was faster deployment. It is generally argued that the old-fashioned tactics of frontal assault by close-packed infantry formations combined with the much-increased effectiveness of rifled muskets to produce the horrendous casualties of the Civil War.

This view has recently been challenged, at least in part, following a careful consideration of the poor training and lack of professionalism of the raw recruits who were swept into the armies of both combatants in large numbers. In this view, much of the superiority of the rifled muskets was wasted: most shots went over the heads of the intended targets because troops did not use the sights properly, and the long effective range

8. For example, regular soldiers of the British Army were trained to hit a man-sized target at 600 yards with their 1853 Enfields.

was compromised by poor visibility much of the time. Only the more sea-soned troops—which included many of the skirmishers—fully exploited the rifled muskets' capabilities.[9]

The effectiveness of the new weapons was felt enough, however, to cause tactics to change over time. Infantry formations loosened up, with infantrymen going into battle more like skirmishers. Artillery pulled back behind the infantry (gun crews and horses made large targets that could be hit from half a mile away). Cavalry did not form into a dense mass and charge infantry across open ground, as they had during the Napoleonic Wars. Indeed, both artillery and cavalry were relegated to subordinate roles during the Civil War; infantry dominated the battlefields. The war became defensive, as frontal assaults were very costly and often failed.[10]

AUTOMATIC WEAPONS BURST

From repeating rifles and double-action revolvers, it is a short step to automatic weapons. Clever means of harnessing the *recoil* energy, or the energy remaining in propellant gas after a bullet had left the barrel, were adopted to mechanize the loading and firing process. A spent cartridge case could be ejected, and a new one loaded into the breech and fired, with increasing rapidity. The first attempt at such a "machine" gun was the Belgian Mitrailleuse, a heavy, multibarreled beast with an ammunition rack that loaded all the barrels together. The barrels were fired by the gunner rotating a lever—so this weapon was not really automatic, though a skilled gunner could get off 100 rounds a minute. Ten years later, in 1861, Richard Gatling invented a multibarreled gun that proved more successful. Gatling's weapon also had a hand crank, but this was used to rotate the six barrels, not the firing mechanism. The Gatling gun, popular with the U.S. Army for decades, could fire 200 rounds per minute. Why did both these weapons have multiple barrels? Because of the rapid rate of fire a single barrel would overheat; multiple barrels would spread the heat and allow a given barrel time to cool down between shots.

9. For an appreciation of the influence of rifled muskets upon the nature of the Civil War, see McPherson (1988, pp. 474–77) and Heidler and Heidler (2002, pp. 1916–17.). A dissenting view is given by Hess (2008).

10. Part of the reason for the dominance of defense is that inexperienced of-ficers, even when they assaulted their enemy successfully, were unable to exploit their success.

Figure 3.3. A British Royal Navy Maxim gun in operation. Note the water jacket around the barrel, and the ammunition belt. Image from Wikipedia.

The first true self-actuated machine gun was the 1883 weapon of Hiram Maxim (fig. 3.3).[11] Using recoil energy to reload, the Maxim gun could fire 600 rounds per minute, with the trigger held down. Its single barrel was cooled using a novel water-filled jacket (the water would boil unless replaced frequently). Machine guns developed rapidly between the 1880s and World War I, setting the stage for the carnage of that epochal event. Once again, weapons technology advanced faster than military thinking. The French, who quickly took up the Mitrailleuse for their war against Prussia in 1870–71, were slow to understand its potential and so did not

11. Machine guns were an American development. Gatling, Maxim, Hotchkiss, Lewis, and Browning were all Americans, though many had to go abroad to see their machines into production. Thus, the Maxim gun is often considered to be a British weapon because it was developed there. Similarly, Hotchkiss went to Belgium and France. Ironically, as a consequence the European powers were ahead of the United States in machine gun development and deployment by World War I. This was also in part because of the U.S. Army's fascination with the less advanced Gatling gun.

Figure 3.4. (a) Soviet 7.62×39 mm AK-47 assault rifle. U.S. Government image. Photographer: Cpl. D. A. Haynes. (b) A British SA-80 L85A bullpup rifle, with a 20-inch barrel and a SUSAT 4× optical sight. U.S. Army photo.

benefit from their advantage. (The Prussians had no such weapon.) By World War I both sides appreciated the machine gun's potential but were slow to develop tactics to counter these automatic weapons.

By the 1890s, all the key components of true automatic firing were understood. The new technology allowed European powers and America to expand the territory under their control, against preindustrialized peoples. As the Anglo-French writer Hilaire Belloc wrote at the end of the nineteenth century, "Whatever happens we have got / the Maxim gun and they have not."

So-called *semiautomatic pistols* were invented by Hugo Schmeisser in Germany in 1916. The trigger was pulled for each round fired, but the hammer did not need to be cocked each time. Thus, semiautomatic pistols were functionally similar to double-action revolvers, with the big difference that the action was powered by recoil and gas pressure instead of by the trigger finger. The mechanism adopted—called *blowback*—was suitable for rapid firing of short-range weapons with pistol ammunition. Similar mechanisms were used for later machine pistols, such as the German MP38 and MP40 (iconic machine pistols of World War II), the Thompson (Tommy) gun, the Russian PPSH41 with its round magazine, and the mass-produced British Sten gun.

World War II saw the introduction of the StG machine gun, a fully automatic hand-held firearm notable as the ancestor of the modern *assault rifle*. Assault rifles have selective fire capability: the shooter can choose to fire a single shot or a burst. Sacrificing the range of a traditional rifle for the rapid fire of a machine gun, the early versions fired pistol rounds. The most famous assault rifle is the AK-47—the Automatic Kalashnikov first introduced in 1947 (fig. 3.4a). This is the world's most popular firearm: somewhere between 30 million and 100 million have been produced. Another iconic assault rifle is Colt's M16; perhaps always in the AK-47's shadow, this American weapon has lasted over 40 years and exists in many versions. The most recent assault rifles, dating from the mid-1980s, adopt the *bullpup* configuration, with magazine and firing mechanism behind the trigger (see fig. 3.4b). Because bullpups have the same barrel lengths as ordinary assault rifles, they have the same accuracy (in principle), yet they are much shorter weapons and therefore easier to carry and use. This trick is achieved by pushing part of the barrel back into the stock.

MUNITIONS

Ammunition development did not cease with the arrival of metal cartridges; much tinkering has resulted in a wide variety of firearms munitions, often highly specialized. The earliest self-contained cartridges employed a *pinfire* ignition system; the primer was ignited by striking a metal pin that protruded from the cartridge. Pinfire cartridges were succeeded by *rimfire* cartridges, in which primer is located around the rim at the base of the cartridge. Today the rimfire system is common only in low-caliber ammunition such as .22 caliber rounds. Two *centerfire* cartridge ignition systems developed in 1866 account for most ammunition today. Named after their inventors, and differing only slightly, the Boxer system is more popular in the United States and the Berdan system in Europe. (A common irony in firearms development: Boxer was an Englishman and Berdan an American.) The Boxer system is more difficult to manufacture but easier to reload (that is, to reuse the brass cartridge case).

The material used to construct cartridges has also developed and specialized. Pistol bullets are now of lead-antimony alloy encased in a soft brass or copper-plated steel jacket, whereas rifle and machine gun bullets have a soft lead core that is encased in a harder steel or cupronickel jacket. Armor-piercing bullets have a hardened steel core. Expanding bullets (developed at the Dum Dum Arsenal in India in the nineteenth century, and banned in warfare) have a particularly soft core that deforms and expands upon impact. Much the same effect is attained by jacketed hollow-point bullets.

Modern *high-explosive* (HE) artillery shells consist of a shell casing, a propellant charge, and a bursting charge. As with small arms weapons, a primer at the base ignites the propellant charge; the bursting charge is ignited by a *fuze* in the nose. Fuzes vary with function: point detonation fuzes set off the bursting charge upon impact; delay fuzes wait a short time before detonation (so that, for example, an antitank round can get inside the tank before exploding); proximity fuzes explode in the air after a set time or at a set distance from the target. (The trench warfare of World War I was a spur to air-burst shell development and to the howitzers that fired them.) Heavy artillery rounds require a *driving band* (also known as a *rotating band*) and a *bourrelet*—two raised rings that are the only points of contact with the barrel. Without these bands the wear and tear upon the barrel would be excessive because of the hard casing of these munitions. The rotating band, near the cartridge base, is grooved to engage the barrel rifling. As well as reducing friction, these bands act as propellant gas seals.

ORDNANCE CATCHES UP

Artillery developed more slowly than small arms for technical reasons: rifling was initially less successfully applied to artillery weapons, and advances in metallurgy were needed before artillery could benefit from smokeless powder.[12] While rifling became standard for small arms during the course of the American Civil War, ordnance remained smoothbore.

12. Artillery shells are much more massive than rifle bullets, with greater inertia; the forces involved in setting a shell spinning are thus much greater. This made it harder to develop rifling that works for artillery. Scale effects (technical note 6) show that powerful smokeless powder could break an artillery gun barrel. This problem was solved not just by advances in metallurgy but also by changing the smokeless powder. Double-base powder gave way to triple-base powder, to achieve the desired pressure curves for ordnance powder ignition.

These problems were overcome in several steps:

- Rifling of artillery barrels became standard in the 1870s.
- The introduction of the interrupted screw breech solved the gas leak-age problem—a more severe problem for large-bore weapons—allow-ing the development of effective breech-loading artillery in the 1880s.
- Cased ammunition (analogous to metal-jacketed rifle cartridges) was developed.
- Improvements in barrel construction permitted powerful smokeless powder charges.
- Pneumatic recoil dampers solved the substantial recoil problems of artillery. With recoil dampers, the barrel slides back on damped roll-ers; this extends weapon lifetime and mitigates gun-laying problems.
- Self-contained firing mechanisms and modern aiming sights were introduced.

The first artillery weapon to benefit from these late-nineteenth-century developments was the famous French 75, which first rolled off the pro-duction lines in 1897. This weapon was a mainstay of the French army in World War I (fig. 3.5a): 75s fired over 200 million rounds during that conflict. The nickel-steel barrel was good for 6,000 rounds; the in-terrupted screw breech provided an effective seal; high-explosive shells packed a punch beyond the capabilities of old-fashioned solid shot or case shot; a trained crew could fire 30 aimed shots per minute.

By the time of World War I, the eclipse of artillery by small arms was a thing of the past. The deadlock of that conflict was due in large measure to the deadly new firepower: long-range artillery and machine guns com-bined (with a lack of mobility, a lack of communications, and a lack of leadership) to produce a static killing ground. As well as being far bloodier than any previous war, this conflict was fought continuously, at a furi-ous pace, right from the start. Artillery was at the fore of any offensive (fig. 3.5b): enemy resolve and numbers were diminished by carefully coordinated rolling barrages that advanced just a few hundred yards in front of the infantry (in theory). Resembling a giant, prolonged siege, World War I has been called "the catastrophe of industrial war."[13]

13. See Hacker (2006) for artillery development during this period, and for the quotation. Marshall (1987) provides a readable account of World War I, and Wolff (1980) gives a telling account of its leadership failures.

(a)

(b)

Figure 3.5. (a) The famous French 75. This gun, introduced in 1897, was the first modern artillery weapon. Here is an anti-aircraft variant, in use on the Salonika front in World War I. (b) A World War I photo of the British 39th Siege Battery in action during the Battle of the Somme, August 1916. The guns are 8-inch howitzers.

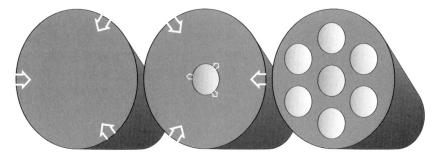

Figure 3.6. Propellant grain shape controls burn rate because gunpowder burns only on the surface. A solid cylindrical grain (*left*) burns degressively, which is to say it burns fast initially and then slows. A cylinder with one hole (*center*) burns out as well as in; the area burning does not change with time, and so the burn rate is constant—a neutral burn. A multiperforated cylinder (*right*) burns with increasing speed—a progressive burn. The timescale depends upon grain size.

Warfare in the twentieth century led to understanding of how internal ballistics works. Gun barrels are designed to withstand a fixed maximum gas pressure; carefully matching barrels with especially formulated double- and triple-base powders led to reliable, accurate, long-range, fast-firing artillery weapons. Recall that smokeless powder grains can readily be formed into desired shapes—important for deflagration rate because powder burns only at the surface. The traditional solid grain—for example, a solid cylindrical grain such as that shown in figure 3.6—produces a *degressive burn*: the burning surface (and thus the rate of gas generation) decreases with time. A cylinder with a single hole in it produces a *neutral burn*: the burn rate is constant. A cylinder with more than one hole produces a *progressive burn*: the area of burning surface increases with time. These grains are also shown in figure 3.6. If we plot the propellant gas pressure as a function of projectile position along the barrel, a degressive burn propellant produces a pressure peak early on, whereas the pressure peak for a progressive burn happens when the projectile is closer to the muzzle and that for a neutral burn is in between these two cases. With the careful formulation of grain composition and shape, a charge can be created that is close to optimum for a given gun (recall technical note 10).

It is possible to investigate the effect of different burn rates mathematically. In technical note 11, our final look at the dynamics of bullets inside the gun barrel, I investigate the relationship between optimum burn rate and barrel length. A mathematical function representing propellant gas

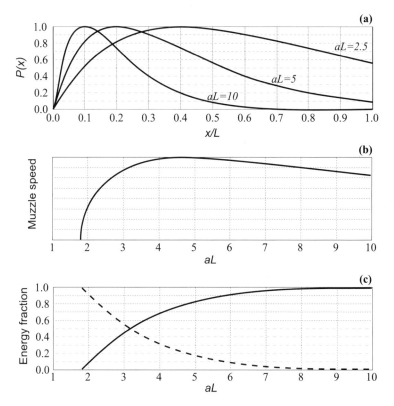

Figure 3.7. (a) Propellant gas pressure vs. projectile position along the barrel. The position of the pressure peak can be chosen, for a given caliber and weight of projectile, by careful choice of grain shape and size. (b) Muzzle speed vs. aL for a simple model of propellant gas pressure (with propellant parameter a; see technical note 11) inside a barrel of length L. There is an optimum choice ($aL = 5$), which shows that propellant burn rate and barrel length must be carefully matched for optimum performance. (c) For the simple model, efficiency (unbroken line) and the fraction of energy needed to overcome barrel friction (dashed line) vs. aL.

pressure is controlled by varying a parameter, leading to pressure curves that correspond to different burn rates, from degressive through progressive. Examples are shown in figure 3.7a. The muzzle speed produced by these different curves is plotted in figure 3.7b. Note that there is a best value, a maximum muzzle speed. This fact shows that we must be careful when choosing grain characteristics; they must be precisely matched with a particular gun caliber and barrel length in order to obtain optimum performance from the gun. (Fig. 3.7c shows the efficiency—solid line—

and the fraction of energy used to overcome barrel friction, for the model of technical note 11.)

Artillery became more mobile between the two world wars.[14] To emphasize the progress that has been made in artillery development during the past century or so, consider the artillery firing tactic known as MRSI (*multiple round simultaneous impact*). It was just about technically feasible in World War II to coordinate ordnance so that different long-range guns in a battery could be fired at slightly different times, in such a way that all their rounds landed in the target area at the same time. That is, guns at different ranges from the target could fire in a coordinated manner so that the target was hit simultaneously by all the guns. This ploy has an obvious tactical advantage: the enemy would have no time to get out of the way before the next round arrived. Today MRSI is more sophisticated: a *single* artillery piece operating at long range can fire several rounds, one after the other, so that these rounds can arrive at the target simultaneously. To achieve this trick, the gunner must load the round with different weights of powder, aim the gun with slightly different elevations, and fire with precise timing. Modern computer-controlled gun-laying can achieve this level of coordination; I show how it is done in technical note 12.

RECOIL

Newton's third law: For every action there is an equal and opposite reaction. Does the bullet fly from the rifle or the rifle from the bullet? Both, of course, and in opposite directions. Rifle and bullet (and shooter) are initially at rest; pulling the trigger releases chemical energy into the system, resulting in the bullet exiting stage right, with momentum mv where m is bullet mass and v is muzzle speed. Newton's third tells us that the rest of the system (rifle and shooter) must acquire the same momentum in the opposite direction—stage left. For a high-caliber rifle with a high muzzle speed, this *recoil*, or "kick," can be significant.[15] For a low-powered rifle or handgun it is negligible; for ordnance it can be huge.

14. For the development of artillery in World War II, see, e.g., Hogg (1970).

15. There are a number of sad and amusing YouTube videos showing shooters who underestimate the recoil of a heavy-caliber weapon. See YouTube (2007e) for video showing the recoil of a .577 rifle firing a 750 grain bullet. (Muzzle speed is 2,460 ft/s for this weapon, and the rifle weighs 13 lb, so that the rifle recoil speed is 20 ft/s.)

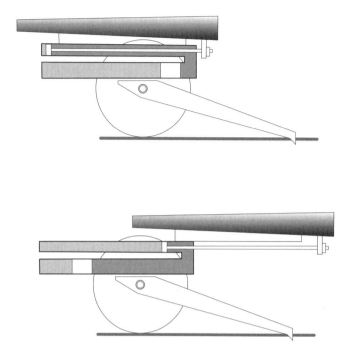

Figure 3.8. Artillery recoil is lessened by permitting the barrel to slide backwards during recoil (*bottom*), with the motion dampened by pneumatic or hydraulic pressure. In the system shown here, liquid (dark shading) compresses gas (light shading) during recoil; the gas then expands to return the barrel to its firing position (*top*).

Recoil was unavoidable until the 1860s, when various mechanisms to mitigate its effects were tried. One of the earliest ideas was to place the gun on a slope. Recoil would push the gun up the slope, but then gravity would return it "to battery"—that is, back to its original placement. Mobile artillery developed a pneumatic recoil mechanism that became commonplace by World War I; the idea is illustrated in figure 3.8.

In technical note 13, I discuss some technical aspects of recoil physics. Here, we can apply some of the derived equations to see how much of an effect recoil has for a given weapon type.

Age of Sail cannon. A long gun from a Napoleonic-era ship might weigh 3 tons and fire a 24-pound ball. Sir Isaac Newton tells us (see technical note 13) that if the muzzle speed is 1,000 ft/s, then the recoil speed of the gun is about 4 ft/s. These cannon were on wheels to facilitate aiming (recall fig. 2.5), and so it was vitally important to ensure that enough strong ropes were provided to prevent the gun breaking free when recoiling. (On a pitching deck, such a "loose cannon" could cause a great deal of damage.)

Heavy rifle. Some modern rifles generate muzzle energies exceeding 10,000 foot-pounds (13,560 J); technical note 13 shows that recoil typically absorbs about 1% of this value, corresponding to 100 foot-pounds of recoil energy—enough to knock a man over (20 ft-lb is considered uncomfortable). Powerful sniper rifles need to be heavy or tied to a frame or have the shooter in a prone position to avoid muzzle rise due to recoil; see figure 3.9. The effects of muzzle rise upon aim are estimated in technical note 13 for a handgun. An experienced shooter who is obliged to fire a large-caliber rifle from a standing position will lean into the shot. This braces him against recoil. (For example, if a 180-pound shooter leans forward enough to lower his height by 1 inch, then 15 foot-pounds of recoil energy will be absorbed just in raising him to his upright position.)

Muzzle brakes. For automatic weapons and large ordnance, recoil is significant, and muzzle brakes are common. A muzzle brake is a series of baffles or diagonal holes cut into the end of a barrel that redirects propellant gas sideways (to reduce recoil) or upward (to reduce muzzle rise). The calculations in technical note 13 show that a long-barrel, low-profile artillery piece that is, roughly speaking, four times as long as it is high would need to redirect about a quarter of the mass of propellant gas, to counter the muzzle rise.

Figure 3.9. The shooter must be careful to avoid muzzle rise when firing this .50 caliber Barrett sniper rifle. It is heavy, which helps, and the stock is in line with the barrel so that recoil induces no upward kick. U.S. Navy photo by J. Husman.

Figure 3.10. An M-198 155 mm howitzer live-fire exercise. The split-trail carriage (rear "legs" of the gun) bury into the ground to absorb recoil; a large muzzle brake redirects the blast. U.S. Army photo by Spc. Lucas T. Shihart.

The trend in artillery over recent decades toward greater mobility (requiring reduced weight and lighter construction) and increased muzzle speed have exacerbated recoil problems. Launcher design is greatly influenced by recoil considerations. Often, a gun design is such that muzzle rise is not a problem, but linear recoil (straight back) is a big problem. In figure 3.10, for example, a howitzer is fired at a sufficiently high elevation angle that there is no muzzle rise. Clearly, however, this gun has a muzzle brake that redirects propellant gas sideways, reducing recoil. The considerable recoil of heavy artillery can reduce the service life of the gun and makes re-aiming at a target more time-consuming, so anything that reduces recoil is beneficial. Note the symmetry of the redirected blast: because the sideways forces cancel, there is no torque acting to twist the barrel. Muzzle brakes can reduce recoil significantly.[16] There is a cost to muzzle brakes, however, and one of these is evident in the figure. Sound pressure levels are higher behind the gun than would be the case without a muzzle brake, and artillery crew must take precautions to avoid hearing injury.

16. For example, the muzzle brake on the sniper rifle shown in fig. 3.9 reduces recoil by 40%. See USMC (2007).

The last logical stage of recoil reduction was the recoilless rifle, developed in the United States in 1963 by William Kroeger and Walter Musser. Propellant gas boosts the projectile forward, and itself backwards out of the breech, so that there is very little gun recoil. Because the weapon is not subjected to large recoil forces, it can be of lighter construction than other ordnance of the same caliber. Two major disadvantages immediately leap to mind: first, the amount of propellant that is required for a given projectile weight is large (between 2½ and 3 times that normally needed); second, the back blast is considerable (for example, when firing a U.S. Army 57 mm rifle, personnel must steer clear of a cone 55 feet long and 45 feet wide in back of the breech).

In the late 1800s two inventions—smokeless powder and the Minié ball—led to a revolution in the design and effectiveness of firearms, first in small arms and then in ordnance. Improved manufacturing techniques were also crucial to these developments. Breech-loading rifled firearms took over from muzzle-loading smoothbore weapons. The accuracy, range, and rate of fire of gunpowder weapons all increased dramatically. Fully automatic weapons were developed during the first decade of the twentieth century.

Part II

WHIZZ!
External Ballistics

4 Short-Range Trajectories
Elementary Aerodynamics

Internal ballistics takes place in the gun barrel; transitional ballistics happens at the muzzle; and external ballistics covers the trajectory from near the gun barrel to the target. So far in this book we have covered the path of a projectile only for a few inches or feet—while it is in physical contact with the launcher, be it a human hand, a catapult sling, or a gun barrel. Most of the remainder of the book will cover the much greater distance of a projectile's airborne trajectory as it describes a graceful arc across the sky. During this phase, a projectile is acted on by the force of gravity and by the force of flowing air—producing aerodynamic drag and sometimes aerodynamic lift. The physics gets increasingly complex as range increases; I will carefully unpack the science of external ballistics over the next three chapters.

TRANSITIONAL BALLISTICS

First, though, there is a gap that must be covered, an awkward transition region in the trajectory of a gunpowder weapon projectile: the physical space that exists between the two regions covered by internal and external ballistics. For muscle-powered weapons such as the longbow or sling, there is no intermediate region; internal ballistics becomes external ballistics. Not so for gunpowder weapons. *Transitional* or *intermediate ballistics* seeks to explain the physics of this transition, as the projectile leaves the muzzle and enters the atmosphere. It is clear where, physically, internal ballistics ends: at the muzzle of our pistol or howitzer, at the instant that the projectile separates from the barrel. It is less clear where external ballistics takes over. For a millisecond or so after it leaves the gun barrel, propellant gas rushes past the projectile, freed from the barrel like gas from a popped champagne bottle. Propellant gas is usually traveling at *supersonic* speed; shock waves both precede and follow the projectile.

Figure 4.1. Ka-BOOM! A full broadside from the USS *Iowa*'s nine 16-inch guns (and six 5-inch guns) displays the power of the blast on the water surface. Impressive—but to a ballistician this represents wasted energy. U.S. Navy photo.

These shock waves can be very damaging to nearby objects—such as gun crews. Aerodynamic drag is not yet a factor that influences the projectile.

In technical note 14, I provide a rough estimate of the physical extent of this awkward and very complicated transitional region by making the reasonable assumption that external ballistics takes over when the propellant gas pressure is reduced to ambient atmospheric pressure. When the projectile emerges from the muzzle, gas pressure is several hundred or thousand atmospheres; the energy of this gas is then dissipated through the air, as is shown dramatically in figure 4.1. Before dissipating, the gas adds a final boost to the projectile, which is now free of muzzle friction. It is important that the rear end of the bullet or shell be symmetric, and that the barrel end be symmetric, so that the blast does not cause the projectile to tumble or to deviate from its trajectory, as this would cause loss of range and accuracy.

Along with gas, unburned propellant is ejected from the barrel, and this propellant ignites in the presence of atmospheric oxygen to produce a flash. Such flashes are undesirable in a military context, particularly for ordnance, because they give away a gun battery's position. Flash suppressors at the end of the gun barrel subdue the flash by causing the propellant gas flow to become turbulent. This turbulence reduces combustion efficiency and so reduces the size and brightness of the flash.

UNDERSTANDING TRAJECTORY PHYSICS: A SLOW LEARNING PROCESS

Since the invention of firearms, people have tried to learn about the shape of the projectile trajectory and understand the forces that define it. For 300 years after gunpowder weapons first appeared, it was thought that the trajectory of the projectile consisted of straight lines connected by circular arcs. In the middle of the sixteenth century the Italian mathematician Tartaglia proved that this could not be the case—that all trajectories must be curved—and that flatter trajectories resulted from projectiles with higher speeds. In the seventeenth century Galileo showed us that if we could ignore the effects of the air, trajectories were parabolic; and Newton showed us why this was the case (gravitational force). Newton was also able to show that air would exert a drag on any projectile and that this drag force would increase as the square of projectile speed. He was unable to solve the problem of drag, however; this was too tough a nut to crack, and another couple of centuries passed before drag effects were understood.

In the next century Benjamin Robins demonstrated experimentally that drag exists: using his ballistic pendulum he showed that the speed of a musket ball decreased with range. Robins also uncovered another aspect of aerodynamics that would not be understood until the nineteenth century. He bent a musket barrel slightly to the right and looked at the trajectories of musket balls fired from this barrel. By placing paper screens at different distances he showed that the trajectory curved to the left.

From the second half of the nineteenth century both experimental and theoretical progress accelerated. Improved chronographs were able to accurately measure the time intervals of projectiles passing specified points in their trajectories; later, high-speed stroboscopic photography told ballisticians just what a bullet was doing as it flew through the air. Some of these movements were unexpected and surprising, as we will see in the next chapter. Ballisticians and aerodynamicists made great theoretical progress in understanding the forces that arose when projectiles interacted with air: aerodynamic drag and lift (both discussed in this chapter), the Magnus force responsible for Robins' bent-barrel trajectories and the twist force arising from rifling (both discussed in chapter 5), and additional smaller forces.

Other physical phenomena that influenced long-range trajectories were also beginning to be understood at this time (say, the half century straddling the year 1900): shock waves and the effect of passing through the sound barrier; earth curvature effects and the *Coriolis force* resulting from the earth's rotation. The development of computers permits us to number-crunch the equations that arise from ballistics analysis with unprecedented accuracy.[1] We can, for example, investigate the influence of atmospheric thinning with altitude upon the trajectories of high-flying artillery shells (a problem that, like the Coriolis force, first became relevant to ballisticians during World War I because of the increasing range of artillery shells). Even so, ballisticians do not yet completely understand *all* the physical effects that have been observed in long-range bullet trajectories. We will see why this is the case in the next chapter. In this chapter I will begin my explanation of external ballistics with the contribution of Galileo and Newton, and then move on to the most significant of the aerodynamic forces: drag and lift.

EXTERNAL BALLISTICS 101: PARABOLIC TRAJECTORIES

All explanations of ballistics must begin with the entry-level calculation of ballistic trajectories in a vacuum. That is, we ignore the effects of air and pretend for a moment that we are firing our gun or throwing our javelin on the moon—or rather, that we are launching our projectile on an earth that has had all the air removed.[2] The reason this is relevant to ballistics on earth (with its dense atmosphere and attendant aerodynamic forces) is that the vacuum trajectory provides an indication of the real trajectory. It is an approximation instead of the true trajectory, but the approximation can be pretty good, at least for heavy, slow projectiles.

Without aerodynamic effects, the only force that acts upon a projectile, once it has been released from its launcher, is gravity. The trajectory

---

1. Much of ballistics falls into a branch of physics called computational fluid dynamics, the equations of which are notoriously difficult to solve.

2. Trajectories on the moon and on the imagined earth-with-no-atmosphere are different because the force of gravity is different.

calculation is relatively simple and is shown in technical note 15. The analysis of technical note 15 is as simple as I can make it without losing some of the physics; when we include real-world aerodynamic effects, the analysis gets harder because more and more complications arise. These effects will be introduced to you in stages. As a general rule you can assume that short-range trajectories are a piece of cake when it comes to understanding the physics. In an analysis of medium-range trajectories we need to take into account more physics, and this seriously complicates things. For long and very long ranges even more physical effects make their presence felt—effects that can be neglected for shorter-range trajectories. I will unpack these effects, and the resulting complications in calculation, in stages over the next couple of chapters.

Here, we get on the escalator with the simple trajectory calculation of technical note 15. We see that if gravity is the only force, then the trajectories are parabolas. (This is not true at very long ranges, as we will see, because earth curvature effects must be included.) Examples are shown in figure 4.2, where I have chosen a launch speed that is appropriate for an arrow. You may look at this figure and think, "Yikes! 700 yards is a long range for an arrow!"—and you are right. Remember that I have removed the entire atmosphere for you, to simplify the trajectory calculations (everybody, please hold your breath until the next section). Arrows flying through air will not achieve ranges anywhere near as long as 700 yards (but we will see surprising exceptions later in this chapter). The maximum range increases as the square of launch speed; it also increases with launch angle up to 45° and then decreases for higher angles. Maximum range increases as the force of gravity is reduced. So an arrow on the moon really might be fired for 700 yards, or further. (Closer to home, an arrow fired on the equator will travel about 0.2% further than an arrow fired in New York or Paris, owing to the earth's equatorial bulge.)

A heavy, slow projectile, such as a javelin, will not be as much influenced as arrows by aerodynamic effects, and so we can expect that the trajectories of javelins will closely resemble the parabolas of figure 4.2. Bullet trajectories—at least those of short range—are not too different from parabolic, as we will soon see. There are two aspects of trajectories that I need to outline here before we move on to tackle aerodynamic effects, and these aspects can be illustrated by appealing to the javelins and bullets just mentioned. First, it is only the projectile's *center of gravity* (CG) that follows a parabolic curve; other parts of the projectile may deviate from a parabola as the projectile turns in flight. Consider a javelin. In

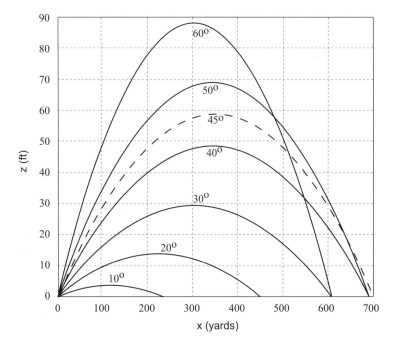

Figure 4.2. Parabolic trajectories for launch angles, θ, of $10°-60°$, with a launch speed of 150 ft/s. Maximum range occurs for $\theta = 45°$ (dashed line). The range for $\theta = 30°$ is the same as for $60°$; similarly, the range for $\theta = 40°$ is the same as for $50°$.

Olympic competition the competitor is shooting for distance; her javelin is launched skyward and with the javelin shaft pointing in the same direction as the initial velocity (to minimize drag, as we will see). Olympic javelins are designed not to turn over in flight (to gain from aerodynamic lift), and so they often hit the ground flat, or even back end first, because the orientation does not change much in flight.

A javelin used in ancient warfare would not fly like this; the heavy point would cause the shaft to turn over during flight, so that the point would stick into the ground (or an enemy) upon landing. The javelin's CG follows a parabola, but the path of the tip, or of the trailing end of the shaft, deviates from a parabola as the projectile rotates in flight. You can see that such javelins must be weighted carefully so that they are light enough to throw but heavy enough at the tip so that they land point-first.

The second aspect of parabolic trajectories (indeed, of all ballistic trajectories, but most easily demonstrated for the parabolic) is that a bullet will drop in flight. Consequently, gun sights must accommodate the effects of gravity. Suppose that you have a hunting rifle with a high sight—let us say that the rear sight is 2 inches above the barrel—and you want to be sure that the bullet trajectory does not deviate from your line of sight by more than 2 inches. What is the maximum range, R, for which this is true? The problem is set up in figure 4.3a.

The solution is provided in technical note 16, equation (N16.2). The maximum R is about 160 yards for a bullet muzzle speed of 1,650 ft/s. R increases as muzzle speed increases, and increases more slowly as d (defined in fig. 4.3) increases; the general formula is given in technical note 16. Suppose now that you want to shoot at longer range. In this case the parabolic trajectory means that it is not possible to sight your rifle so that d is less than 2 inches for all ranges between you and the target. Instead, d is within the allowed deviation over a smaller range swath, as illustrated in figure 4.3b. You adjust your sights for the range desired; in technical note 16, I calculate the extent of the range swath (from approximately $R - r$ to $R + r$) for which the trajectory deviates by less than 2 inches from the line of sight. For the same muzzle speed, and a range of $R = 250$ yards, we have $r = 38$ feet. In other words, the range swath for which the sight setting works is about 10% of the range (i.e., $2r/R = 0.1$). In technical note 16 the

(a)

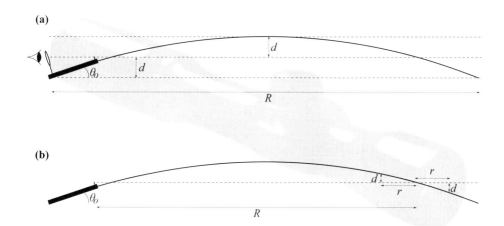

(b)

Figure 4.3. (a) Setting high sights so that the parabolic trajectory (curved line) out to range R deviates from the line of sight (middle dashed line) by no more than d. For the example in the text $d = 2$ inches. (b) More commonly we set sights for a particular range R, and the bullet trajectory deviation is less than d for a swath of ranges between $R - r$ to $R + r$. We can calculate r.

general formula for r shows that range swath increases as the square of muzzle speed. So if your rifle had a muzzle speed of 2,000 ft/s, the sight setting works for a swath that is 15% of the range. This shows one of the benefits of high muzzle speed.

Of course, real trajectories are not quite parabolic because of aerodynamic effects, particularly aerodynamic drag. But the results of my simple parabolic calculation are close enough to convey the general ideas:

- In general there are two trajectories, low and high, for a given range and a given muzzle speed.
- Bullet drop means that sight setting works only over a limited range swath. The extent of this range swath increases as the square of muzzle speed.

In figure 4.4 you can see how well the parabolic approximation works for a heavy, low-speed projectile—a seventeenth-century mortar, firing at high angles.

Now it is time to replenish the earth's atmosphere and tackle the consequent aerodynamic effects for ballistic trajectories. You may breathe again.

INTRODUCING AERODYNAMICS: WHAT A DRAG

Aerodynamic drag is the force you feel on your hand when you hold it out the window of a moving car. This everyday experience will tell you two obvious facts about the drag force—obvious but, to a physicist, complicating. First, the drag force must act in the opposite direction to velocity. This complicates the physics because it means that the drag force, unlike gravity, changes direction as the bullet velocity changes direction. Second, the magnitude of the drag force depends upon the shape and size of the projectile: bigger bullets will experience more drag than small ones, streamlined bullets will experience less drag than flat-nosed ones. This complicates analysis because the dependence of drag upon shape is complicated, and because the shape that a bullet presents to the onrushing air—its cross-sectional area—can change over the course of a trajectory, meaning that the drag force also changes.

In technical note 17 I set up the equations of motion for a bullet subjected to two forces: gravity and aerodynamic drag. In general, these equations cannot be solved analytically; we have to resort to a computer. (This facet of ballistics is typical of fluid-dynamical problems in physics.

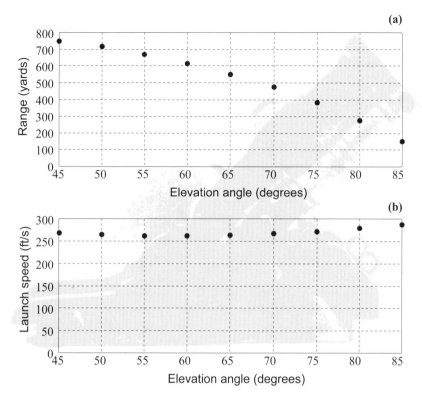

Figure 4.4. (a) Data from firings of a seventeenth-century mortar: range vs. elevation angle. (b) Mortar bomb launch speed, inferred from the parabolic equation. If this equation is exactly right, and if the data is perfect, launch speed should be independent of elevation angle—it is nearly so. The slight variation in speed with elevation looks like systematic error, not random error; it is probably the small effects of aerodynamic drag. Data from Phillippes (1669).

Fluid dynamics is hard.) For the special case of flat trajectories—a special case that applies to most bullet trajectories—it is possible to derive approximate solutions to the equations. These solutions are also presented in technical note 17 and discussed here.

First, though, I need to subject you to some of the myriad complications of drag force physics. The idea is not to befuddle you but instead to indicate why drag is such a source of confusion. I aim to "declutter" the subject by presenting the results of a simplified analysis: simplified, but not oversimplified, to reveal the essential physics of drag forces. The form that we adopt for the drag force is this:

$$F_D = \frac{1}{2} c_D \rho A v^2. \tag{4.1}$$

If you are not following the math in detail, here is the same information in words. The drag force, F_D, depends upon the square of bullet speed, v, as Sir Isaac Newton predicted. Other elementary considerations show that drag force depends upon fluid density, ρ, and upon cross-sectional area, A (i.e., the area of a bullet that you see heading straight at you).[3] These three factors—bullet speed and cross-sectional area, and air density—cover most of the bases, but there are some residual effects that greatly complicate drag force physics. The residual effects are swept up and dumped into one metaphorical bucket labeled "drag coefficient," c_D. That is to say, the drag force is written in terms of bullet speed, bullet area, and air density, as described, but all these factors are then multiplied by a drag coefficient that contains all the extra physics.

In many circumstances it is satisfactory to assume that the drag coefficient maintains a constant value throughout the trajectory of a bullet—this is the simplifying approximation that I exploit in technical note 17—but you should always bear in mind that the resulting predictions about bullet behavior are approximate. The uncomfortable truth is that drag coefficient is not always constant throughout a trajectory. In general it varies, sometimes quite markedly. Same goes for the cross-sectional area, and even sometimes for the air density, as we will see later on. However, for stable flight (again, I must defer a discussion of flight stability and instability until later; we must learn to walk before we run) it is true to say that the drag coefficient variation is small compared with the average value. This fact makes it sensible to approximate the drag coefficient as constant.

One circumstance where the drag coefficient is most certainly *not* constant applies when the bullet breaks through the sound barrier. Most bullets emerge from the gun barrel at supersonic speed but then slow down because of drag. For a long trajectory the speed can drop below the speed of sound, at which point the drag coefficient does a backflip and drops precipitously. This in turn may cause the bullet to do a backflip—quite literally: it tumbles. Here is one example of a situation where we have to be careful about applying standard analysis.

3. The fluid is that through which the bullet travels, usually air. This dependence of drag force upon fluid density is why bullets slow down more quickly in water (because water is so much denser than air).

(a)

(b)

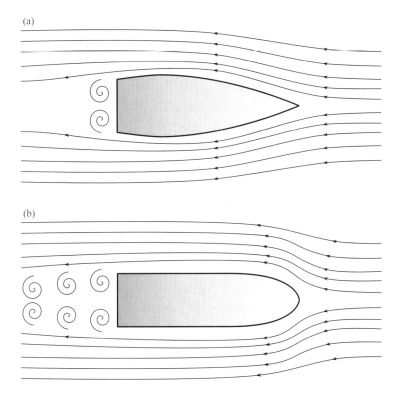

Figure 4.5. (a) A streamlined bullet: sharp point and boat-tailed. (b) Less well streamlined: blunt nose and square rear. This shape generates more turbulence in the wake.

Another complication that arises when we consider drag coefficients is the way that they depend upon bullet shape and orientation. First, let's consider the shape. Figure 4.5 depicts schematic illustrations of two bullets and the streamlines formed around them in the airflow. We can see qualitatively why the pointed, boat-tailed bullet will have a smaller drag coefficient. At the front, the pointed bullet parts the air more slowly than does the blunt bullet and so exerts a smaller force on the air (assuming that the bullets have the same speed). The energy expended by the pointed bullet in moving aside the air is therefore less than the energy expended by the blunt bullet. Since bullet energy is proportional to speed squared, less energy expended means less speed lost—that is, lower deceleration and thus a smaller drag coefficient. At the back, the boat-tailed bullet guides the two central streamlines—parted by the bullet—back together again like a zipper.[4] Ideally, from an aerodynamic viewpoint, the bullet should be pointed at the back as well as at the front because this would zip together the streamlines more effectively, but practical bullets must have a flat area at the back (recall the internal ballistics). Consequently, air immediately behind the bullet becomes turbulent as the airflow separates and forms a wake; this saps energy from the bullet. The problem is worse for the square-ended bullet. In this case the volume of turbulence at the back is greater and the energy lost is greater. So, looking at the bullets from both front and back, we can understand why one of these bullets has a smaller drag coefficient than the other.[5]

The arguments that I have just presented apply pretty well to bullets and other projectiles that are *subsonic*—traveling at speeds below the speed of sound. At supersonic bullet speeds, however, the situation is different. Consider now figure 4.6, where we are looking at the pressure waves generated by a projectile as it plows its way through the air. This is a different aspect of aerodynamics that was not shown explicitly in figure 4.5, where we looked at streamlines. The caption to figure 4.6 shows how shock waves arise. The angle of the shock wave front depends upon bullet speed,[6] not upon the shape of the bullet. The significance of shock waves is that they take away a lot of energy from the bullet. The wave emanating from the bullet nose is one of high pressure, and that at the tail is low pressure. Consequently, the aerodynamic drag of a bullet increases sharply as a projectile accelerates through the sound barrier. (Fig. 4.7 provides a dramatic illustration of the shock wave created as an airplane accelerates through the sound barrier.) The drag coefficient falls at higher speeds. For

4. The phenomenon of streamlines following the shape of gentle curves is known as the *Coanda effect* and is an important consideration in aerodynamics.

5. The low-pressure cavity that forms in the wake of a projectile—part of the phenomenon of tail drag—can be reduced (and tail drag reduced) in artillery shells by making use of *base bleed*. Base bleed refers to the emission of a jet of gas from the base of the shell to fill the cavity. It reduces the drag of a M2 105 mm round by half. Spoilers at the base of some projectiles similarly fill the cavity and reduce tail drag.

Bullets with a pointed nose first appeared in the French Lebel rifle cartridge, in 1898. The bullet was referred to as a *spitzer*, from the German *spitz*, meaning pointed.

6. In fact the angle depends only upon Mach number—the ratio of bullet speed to the speed of sound. It is not difficult to show, from the diagrams of fig. 4.6, that the shock wavefront angle Ω is given by $\sin(\Omega) = u / v$, where u is the speed of sound.

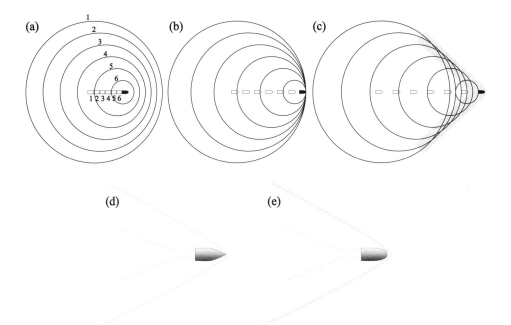

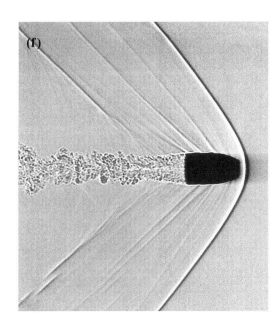

Figure 4.6. (a) At subsonic speeds, a bullet sends out pressure waves, some of which precede the bullet and act as an air cushion. This makes bullet nose shape unimportant so far as drag coefficient is concerned. (b) When the bullet is travelling at the speed of sound, the pressure waves pile up at the nose: this is the sound barrier. (c) At supersonic speeds, there is a shock wave trailing from the bullet. The drag this creates depends sensitively upon bullet nose shape. Air outside the shock wave front (the *Mach cone*) is completely unaffected by the projectile. (d) For a bullet with a pointed nose, the shock wave does not take away as much energy as for a blunt-nosed bullet (e), so the blunt bullet has a higher drag coefficient in the transonic region. (f) Shadow micrograph showing the shock waves from a real bullet.

this reason, gun manufacturers like to make weapons with muzzle speeds that are not close to the speed of sound; they choose either higher or lower speeds, where the drag coefficient is reduced.

These various contributions to aerodynamic drag are given names in the technical literature to emphasize their different physical origins and behavior. Drag due to friction at the projectile surface is called *skin drag*; it acts tangential to the surface. Skin drag is not very important for short-bodied bullets but becomes significant for long-bodied projectiles such as missiles and rockets. *Tail drag*, already discussed, is the dominant drag effect below Mach 1—that is, at subsonic speeds. Drag that is due to pressure differences over the surface is known as *pressure drag* (or *form drag*); it acts perpendicular to the surface. Together, the contributions from tail drag and pressure drag are termed *profile drag*. Pressure differences across a shock wave give rise to a type of drag that appears only at supersonic projectile speeds; this is known as *wave drag*. A fourth kind of drag force, which we will meet later, is *induced drag*; it arises when lift forces cause a downwash of air that spills off the projectile surface as

Figure 4.7. An FA-18 Hornet accelerates through the sound barrier, generating a shock wave. U.S. Navy photograph.

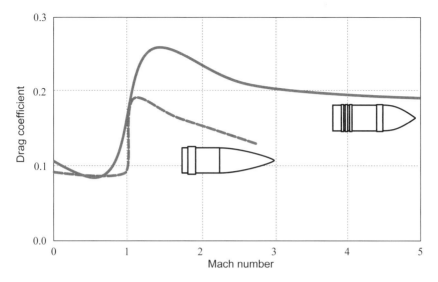

Figure 4.8. Drag coefficient vs. Mach number for two artillery rounds (illustrated). Data from a Canadian field artillery manual (Canada DND 1992).

vortices. These vortices carry away energy and contribute to projectile deceleration.[7]

From figures 4.5 and 4.6 we can construct a picture of how drag coefficient varies with bullet speed and shape. Subsonically the tail region matters—boat-tailed bullets have less drag for reasons already discussed—but the nose shape is unimportant. Why? In figure 4.6a we see that a subsonic bullet sends out a pressure wave in front of it. This air cushion pushes air molecules out of the way; it is as if the bullet is advertising its imminent arrival ("Here I come—get out of my way!"). At supersonic speed, on the other hand, the nose shape is very important. Pressure piles up very close to the nose; there is no wave advertising the bullet's approach—it arrives unannounced. As a consequence, the region of high pressure in front of the bullet depends upon nose shape: the high-pressure area is larger for a snub-nosed bullet than for a pointed bullet. Large high-pressure area produces large drag, so nose shape matters at supersonic speeds. In figure 4.8 you can see the measured drag coefficients for two types of artillery rounds; note the jump at M1.[8]

7. Massey (1989) and Canada DND (1992) provide good technical descriptions of the different contributions to drag forces.

8. Settles (2004) has produced high-speed images showing the shock waves that

So, bullet shape is a compromise that takes into consideration the physics of both subsonic and supersonic regions in order to reduce aerodynamic drag. Tail shape matters subsonically; nose shape is more important supersonically. Yes, this is a complicated subject, and it only gets worse.

At low speeds, the drag coefficient of a projectile depends upon the *Reynolds number*, which is the ratio of viscous to inertial force acting upon a projectile; this important aerodynamic quantity is the dominant parameter when the fluid density does not change much. Such is the case for arrows and javelins. At high speeds the fluid becomes compressed, and the drag coefficient depends upon *Mach number*, which is the ratio of projectile speed to the speed of sound. This is the region that applies to bullets. At intermediate speeds, the drag coefficient varies with both the Reynolds number and the Mach number. It is possible to calculate the drag coefficient in some limiting cases for projectiles of simple shape (for example, for an old-fashioned spherical musket ball at very low speed), but in general the drag coefficient has to be determined experimentally. This is a physicist's way of saying that we don't fully understand the physics of ballistics (because it is so complicated).

Before presenting you with results from the simplified drag force calculation provided in technical note 17, I will briefly summarize what we have learned about the nature of aerodynamic drag. Except near the speed of sound, it is approximately true to say that drag force increases with air density, with projectile cross-sectional area, and with the square of projectile speed. At the level of this approximation (let us denote such a level of analysis as level 1), the drag coefficient c_D of equation (4.1) is a constant, as in technical note 17. To proceed to a more accurate analysis (level 2) that accounts for viscous forces and the effects of air elasticity, we must consider c_D to be a function of the Reynolds number and the Mach number, which depend upon projectile speed in different ways. Level 2 analysis is much more difficult than level 1 and requires heavy-duty computer number crunching. Even level 2 is inexact; we will learn later that the other parameters of equation (4.1), ρ and A, are also variable. Including such effects (level 3) is, with one exception, beyond the scope of this book. I will, however, show you some of the features and results of level 3 in the next chapter.

---

emanate from a bullet as it travels through the air, and the shock waves arising from the muzzle blast.

JARGON TERMS

If you are an active shooter, you must have heard of "ballistic coefficient" and "sectional density." These two terms, commonly applied in North America to describe the aerodynamic properties of bullets, are more spoken of than understood, and so it is worthwhile describing what they mean, even though (for reasons that will become apparent) I will not be using them much in this book.

The *sectional density* of a bullet is its mass (or sometimes weight) divided by its caliber squared, $D = m/c^2$. A bullet's drag deceleration (the reduction of bullet speed due to aerodynamic drag) depends upon its sectional density plus other aerodynamic parameters that are independent of bullet geometry.[9] Or so it used to be thought in the late nineteenth century, when ballisticians were seeking a convenient parameter for describing the aerodynamic properties of a bullet. In other words, people thought that drag deceleration was a constant multiple of sectional density. (This is equivalent to saying that the drag coefficient is unvarying.) The catch: measuring the constant was difficult to do in those days. Doing so required measuring the speed of a bullet at two different points of its trajectory, so this measurement was performed only for a standard reference bullet of 1 inch in diameter weighing1 pound. The idea was that a shooter could then estimate the drag deceleration of *any* bullet by multiplying its sectional density (easily measured) by the constant obtained for the standard bullet.

Reality intervened. Aerodynamic drag is much more complicated than was appreciated at the end of the nineteenth century, as we have seen. It was soon learned that the constant was not constant: it was different for different bullets. To get around this failing, sectional density was replaced by the *ballistic coefficient*, which is the sectional density of a bullet divided by a *form factor*, denoted by i: $B = D / i$. The form factor is the bullet's drag coefficient divided by the drag coefficient of the standard reference bullet: $i = c_D / c_{D0}$. The value assigned to the form factor is determined by measuring it for each type of bullet. Recall that the whole point of

9. We can see that deceleration depends upon sectional density from eq. (4.1). Divide both sides of the equation by projectile mass; the left side of the equation is now projectile acceleration (in practice, deceleration), and the right side of the equation depends inversely upon D (note that cross sectional area A is proportional to caliber squared, c^2).

introducing sectional density was to avoid such measurements. Worse yet: it was later discovered that the ballistic coefficient was not even constant for one bullet: it changed during the bullet's flight as the speed changed. So, nowadays there are several ballistic coefficients that need to be specified in order to give the shooter an idea of how his bullet will perform.

Ballistic coefficients are an anachronism left over from the early days. They are popular with ammunition manufacturers, however, and the numbers are often reported optimistically to boost sales.[10] For square-ended, snub-nosed bullets the ballistic coefficient is low (about 0.1 or 0.2 lb/in^2) whereas for more aerodynamic bullets (such as the boat-tailed bullet with a sharply pointed nose shown in fig. 4.5a) B is larger (perhaps (0.4–0.6 lb/in^2). Very-low-drag bullets have a ballistic coefficient exceeding 1.1 lb/in^2. I will not use ballistic coefficients explicitly in this book but will instead stick with c_D, the aerodynamicist's measure of drag, or with the related drag factor, b_D, introduced in technical note 17.

THE INFLUENCE OF DRAG UPON TRAJECTORY

The muzzle speed of a rifle bullet is typically two or three times the speed of sound in air—in other words, Mach 2 (M2) or M3. For a NATO 7.62 mm ammunition round, for example, the drag coefficient at M1.5 is $c_D \approx$ 0.40, falling to $c_D \approx 0.25$ at M3. So for my level 1 analysis, where we must choose a constant value for drag coefficient, let us say that $c_D = 0.30$. Given this value for the drag coefficient, the analysis of technical note 17 produces short-range bullet trajectories as plotted in figure 4.9 for two different bullets at two elevation angles. Note how the range is shortened as a result of drag force. It is also distorted, compared with the vacuum trajectories (also shown in fig. 4.9); the trajectory angle at landing is bigger than the angle at launch, and the maximum height occurs a little more than halfway along the trajectory.

From technical note 17 we can also calculate the flight time for a bullet to complete a trajectory and the bullet speed at the end. The trajectories of figure 4.9a have flight times of 0.31 s for the M-16 rifle round and 0.33 s for the M80 rifle round.[11] Bullet speed at the end of the trajectory is 2,380 ft/s

---

10. See the online preprint of Courtney and Courtney (2007).

11. I refer here to the Zastava M-80 assault rifle (a Kalashnikov derivative chambered for 7.62×51 mm ammunition) and not the Sauer M-80 hunting rifle.

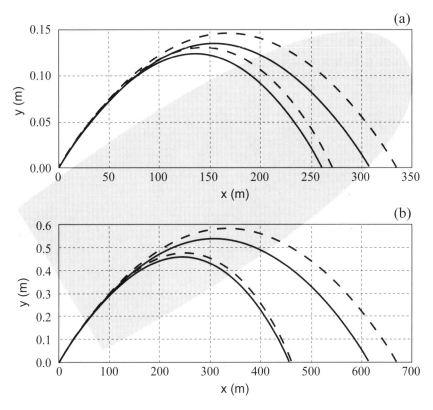

Figure 4.9. Short-range bullet trajectories in a vacuum and in air—i.e., without aerodynamic drag (the higher trajectories) and with drag (the lower ones). Solid lines: M-80 NATO 7.62×51 mm ammunition; dashed lines: M-16 NATO 5.56×45 mm ammunition. (a) Elevation angle $\theta_0 = 0.1°$. (b) $\theta_0 = 0.2°$.

(74% of muzzle speed) for the M16 round and 2,420 ft/s (79%) for the M80 round. For the longer trajectories of figure 4.9b, flight times are both about 0.62 s, and final speed is 60% (M16) and 66% (M80) of muzzle speed. You can see that, even for these short trajectories, the effect of aerodynamic drag upon bullet trajectory and bullet speed is significant; for longer trajectories the effect is magnified.

We can compare the predictions of our level 1 calculation against data from real trajectories. The mathematical model of technical note 17 predicts that bullet speed decreases with distance from the muzzle, as shown in figure 4.10 for three types of rifle ammunition. Also plotted in the figure are the measured data for these rounds. The agreement is pretty good,

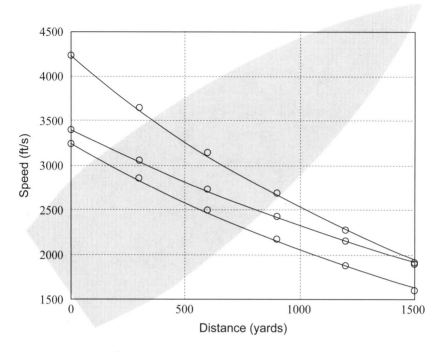

Figure 4.10. Speed (ft/s) vs. distance from muzzle (yards) for three rifle rounds. Open circles are data from Hornady rifle ammunition data sheet; lines are theoretical values calculated from technical note 17. *From top to bottom:* 204 Ruger 32-gr V-MAX ammunition, 243 Win 75-gr HP ammunition, 223 Rem 55-gr V-MAX ammunition.

giving us confidence that this level 1 model captures much of the reality of bullet trajectories.

ADD AERODYNAMIC LIFT

We are all familiar with the concept of lift force from our experience with airplanes: at takeoff we are pushed into our seats as the plane leaves the ground. Because of this experience we tend to think of aerodynamic lift as a force that acts vertically upward, in the opposite direction to gravity. In fact, lift can act in any direction, depending upon the airfoil design and orientation. A sailboat is propelled by the aerodynamic lift provided by wind in the sails. In this case the sail acts like an airplane wing on its end, and consequently the lift force is directed more or less horizontally. The airfoil on a drag racer is shaped to provide a lift force in the downward direction, to increase tire traction with the track surface. Lift is technically very similar to drag; together, they are two components that make up the force of air acting upon a body (in our case, a projectile) moving through it. Let us call this force the *wind force*. Drag is the component of wind force that acts in the opposite direction to the projectile velocity; lift is the component that acts perpendicular to velocity.

Lift is a major factor in some projectile trajectories—for example, those of arrows—but it is less important for others, such as those of bullets. Bullet lift is a chord played in a quite complex symphony of bullet trajectory dynamics, so I will defer discussion of it until the next chapter. Here I will introduce aerodynamic lift (at level 1) by applying it to *flight arrows*, where its application is most clearly evident.

A flight arrow is one that is designed for maximum range. It can be fired from the same bow as other arrows but will have a longer trajectory because the flight arrow weight and shape are chosen to maximize lift force and thus (as we will see) increase range. In fact, a major component of flight arrow performance arises from reducing aerodynamic drag as much as enhancing aerodynamic lift; the parameter that matters is the lift-to-drag ratio. Drag is reduced by making the arrowhead small and pointed, and by constructing the shaft so that it is as short as practicable, is thin and tapered at the ends, and has a center of gravity, or CG, that is forward of center. (To reduce drag we want the arrow shaft to be in approximately the same direction as the arrow velocity vector, which requires a forward CG.) Historically, arrow fletchings are made from feathers, but flight arrow feathers need to be thin so as to reduce drag. (Razor blades have been used for this purpose.) To enhance lift the fletching shape is chosen aerodynamically, and the shaft is lightweight (though not too light because this would enhance drag). The arrow weighs about 220 grains, or half an ounce. The optimum angle between shaft and velocity direction, called the *angle of attack* when referring to an airfoil, should be small (to minimize drag) but not zero: the arrow head is slightly above the tail. This arrangement (achieved by careful choice of CG) ensures that the lift force is directed more or less upward.

Having carefully constructed our flight arrow, we now launch it skyward from as powerful a bow as we can handle. What is the optimum launch angle, given lift and drag forces? What ranges can be attained? The answers to both these questions are surprising. My level 1 analysis is summarized in technical note 18. The lift force assumes exactly the same form as the drag force—that of equation (4.1). This should not be a surprise, given that lift and drag are but two components of the same force. The only difference is that the drag coefficient in equation (4.1) is replaced by a lift coefficient, c_L. The lift coefficient is assumed to be constant in technical note 18 (this is what I mean by "level 1," recall), but in reality it varies. The lift coefficient changes as the angle of attack changes, and so any wobbles of the arrow shaft during flight will lead to a varying lift coefficient. However, we are assuming that our flight arrow has been well designed, in which case it will not wobble much (because wobbles increase drag); thus. we can reasonably assume that lift coefficient is constant during flight.

A medieval longbow firing a heavy war arrow could achieve ranges of about 250–300 yards. Contrast this typical arrow range with the modern world record for flight arrow trajectory length of over three-quarters of a mile (1,222.01 m, to be exact), set in 1987 by Don Brown of the United States.[12] He used a bow with a very heavy draw weight, but even this is not essential for obtaining extraordinary arrow ranges: the current record for a flight arrow fired from a bow with the modest draw weight of 25 kg (55 lb) is over 1,000 yards. Nor are these distances purely the result of modern technology (modern bows and arrows are both very high-tech): a Turkish archer in the sixteenth century shot a flight arrow over 900 yards.[13] We can estimate, using arrow geometry, that a typical arrow might have a drag factor of $b_D = 0.003$ m^{-1}. The drag factor of a flight arrow is much smaller because of its careful construction, but it is difficult imagining it being smaller than $b_D = 0.0005$ m^{-1}, so I will adopt this value in the calculations to follow.

Can the analysis of technical note 18 reproduce the astonishing ranges achieved by flight arrows? This analysis shows that arrows with very low drag are necessary but are probably not enough. Without lift, to achieve a range of 1,200 m the arrow would require a launch speed of about 120 m/s (400 ft/s)—just about possible, but pushing the envelope. I will limit launch speed to 100 m/s, which is certainly possible for a powerful bow. In figure 4.11 you can see the results of computer simulations based upon the calculations of technical note 18: with no lift and a flight arrow with very

12. See FITA (2009) for flight arrow distance records. For typical arrow launch speeds, see, e.g., Lauber (2005, chap. 16).

13. A discussion of this claim, and a simpler analysis, are provided in Denny (2007); see chap. 1 and references therein.

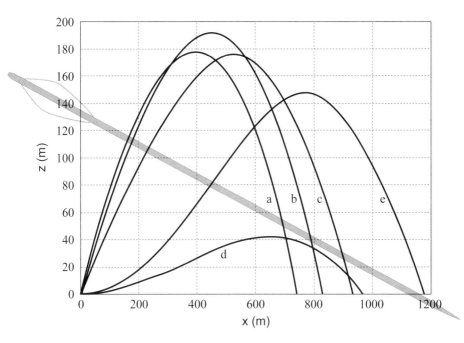

Figure 4.11. Flight arrow trajectories calculated in technical note 18. The arrow's initial speed is 100 m/s and its drag factor is 0.0005 m^{-1} (which corresponds to a drag force of about half the weight of the arrow). If f is the initial lift force, expressed as a fraction of arrow weight, and θ_0 is elevation angle, the trajectories shown are (a) $(f, \theta_0) = (0.0, 40°)$; (b) $(f, \theta_0) = (0.5, 35°)$; (c) $(f, \theta_0) = (1.0, 25°)$; (d) $(f, \theta_0) = (1.5, 0°)$; and (e) $(f, \theta_0) = (2.0, 0°)$. The launch angles are those that maximize range.

low drag, the maximum range is about 730 m.[14] Modest lift forces take the maximum range out to 900 m; to reach 1,200 m, the lift force at arrow launch needs to be about twice the arrow weight. The maximum range is obtained for a ratio of lift force to drag force of about 4 ($L/D = 4$); increasing lift causes the arrow to rise high and then plummet, rather than traveling horizontally. Note from figure 4.11 how the optimum launch angle decreases as lift force increases; at very long ranges the optimum angle is zero. The arrow is fired horizontally and curves up owing to lift— the arrow flies like an airplane—before falling, as drag reduces speed (which reduces lift).

14. For arrow trajectories we need to solve equations (N18.1) in technical note 18. These equations can be solved only by computer number-crunching.

STABILITY

In the next chapter I will share with you the external ballistics of bullets and artillery rounds fired to medium and long ranges. The flight dynamics of these projectiles is complex; one of the complicating features that I will be discussing is projectile stability. Stability refers to the projectile's orientation as it flies through the air. The flight is said to be stable if the orientation is more or less fixed. In stable flight the arrow or bullet is allowed to slowly change orientation so that it is aligned with its trajectory (so that it points along the direction of motion). The flight is deemed to be unstable if the projectile orientation wobbles excessively or if the projectile tumbles end over end. I can broach this subject here because arrow stability is simple to visualize and provides a gentle introduction to the subject.

An arrow head is heavy and aerodynamically streamlined, whereas the fletching at the tail is light and (except for flight arrows) draggy. Consequently, the arrow's CG is forward of the geometrical center, whereas its *center of pressure* (CP) is aft of center. The CP is the point at which aerodynamic forces can be considered to act (just as inertial forces are considered to act at the CG). So, arrows always have their CG in front of their CP. This happy circumstance means that the arrow flight is stable. You can see why this is so. As the arrow flies, if the tail end sags below the arrow head or rises above it, then air catches the fletching and pushes the arrow back into line. If the CP had been in front—say the fletching feathers were at the front—then any movement out of line would be caught by the wind and the arrow would flip 180°. As it flipped, the arrow would present an increased cross-sectional area to the airflow, which, from equation (4.1), would cause the drag force to increase. It would also cause the lift force to disappear. The result of a flip would be an unpredictable trajectory and an inaccurate shot.

The message to take away from this brief introduction to projectile flight stability is that stability is a consequence of projectile shape. If the CG is ahead of the CP, then stability reigns. If, on the other hand, the CG is behind the CP, the projectile is unstable and the trajectory will be unpredictable.

A bullet or shell emerging from the barrel enters a transitional region, where the physics is complex. If aerodynamic effects could be ignored, the ballistic trajectory of a projectile would be a simple parabola. In reality, aerodynamic forces are significant; unfortunately, they are also complicated, particularly when the projectile speed is close to that of sound. Aerodynamic drag changes the trajectory shape and reduces range. Aerodynamic lift for projectiles such as arrows can increase range significantly.

5 Long-Range Trajectories
Advanced Aerodynamics

I ended chapter 4 by observing that a projectile with a center of pressure (CP) forward of its center of gravity (CG) would be aerodynamically unstable. This is precisely the case for our most common ballistic projectiles: bullets (and also artillery shells). Why are these projectiles unstable, and what can be done to stabilize their trajectories? We already know the answer to the second part of this question: bullets and shells are given a spin, via rifling in the barrels. Shortly, we will see how spinning a projectile makes its trajectory stable. There are other aerodynamic consequences of spinning, and I will discuss these throughout the chapter. To see why the CP of an artillery round is in front of its CG, consider its weight distribution. The heavy shell casing and payload—the explosive charge, say, for a high-explosive (HE) round—are at the rear and middle; the front consists of a light, hollow cone (or *ogive*, or elliptical section, or . . . I will summarize the extensive and much-researched topic of projectile nose shape in the next section) for aerodynamic streamlining. Thus, most of the weight is at the back, whereas the central point at which we can assume air pressure acts is near the front.

Bullets do not have a hollow nose (*meplat*). The CG is further forward than it is for shells but is still behind the CP; therefore, bullets are aerodynamically unstable and must be stabilized gyroscopically by imparting a spin to them.[1] The shape of bullets and shells is a trade-off between three

---

1. The meplat is the front surface of a bullet or shell; the term derives from the French *méplat*, meaning "plane," inappropriately enough. (Planes are flat, whereas the front surfaces of bullets are curved.) One exception to the aerodynamic instability of bullets might be the early Minié ball: it was short and had a conical hollow at the back to help the soft metal expand into rifling grooves, as you may recall from chapter 4. This geometry may have been enough to bring the CG in front of the CP.

competing effects: the demands of internal ballistics and of aerodynamics, and the requirements imposed by terminal ballistics.

BULLET SHAPES

I mentioned in chapter 4 how bullet shape influences aerodynamic drag, in a speed-dependent manner. Much research effort has gone into this subject, as you may imagine: bullet, shell, and missile designers would like very much for drag to go away, because it reduces projectile speed, range, and accuracy. Of course, drag does not go away, but its effects can be minimized by judicious choice of projectile shape, particularly the nose cone. A number of nose cone shapes that have been tried and tested are shown in figure 5.1. These shapes are all axially symmetric, which is to say they are formed by drawing a two-dimensional curve (such as a section of a circle) and then rotating the curve about the longitudinal axis of the projectile to define a three-dimensional shape. Because of their simple geometry, such designs are easy to manufacture, in contrast with more theoretical shapes that are not formed from circular or elliptical sections.

Apart from projectile speed, one of the important parameters for nose cone design is the *fineness ratio* (also, most unfortunately, sometimes called the *caliber*, just to confuse us). Fineness ratio is the ratio of nose cone length, L, to radius, R. In 1941 aerodynamically optimum nose cone shapes were derived from theory by Wolfgang Haack.[2] These shapes are shown in figure 5.1e for a fineness ratio of 5. Here, one shape corresponds to minimum drag for a given projectile diameter (bullet caliber), and one shape corresponds to minimum drag for a given bullet weight. As the fineness ratio increases, wave drag (introduced in chapter 4) decreases but skin drag rises, so the optimum choice for fineness ratio is a trade-off; $L/R = 5$ is a common choice.

In the *transonic* region, corresponding to projectile speeds of between M0.8 and M1.2, the drag of nose cone shapes changes markedly with speed. The ogive shapes shown in figure 5.1 tend to perform poorly at transonic speeds, whereas a parabolic shape or the von Kármán nose cone of figure 5.1e does better. Because of the dependence of drag upon speed, it is not difficult to tell the design speed of a bullet from its shape. Stated

2. Wolfgang Haack was a German mathematician who discovered the minimum drag shapes in 1941. His work was kept secret until after World War II. See Boos-Bavnbek and Høyrup (2003) for details.

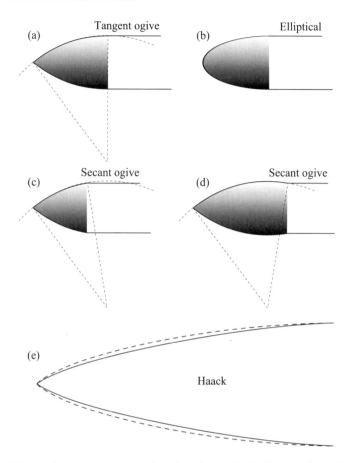

Figure 5.1. Aerodynamic nose cone (meplat) shapes for bullets, shells, and missiles. (a)–(d) Two-dimensional geometrical shapes, rotated about the longitudinal axis to form a nose cone. (e) The optimum Haack nose cone for a fineness ratio of 5.0 (see text). The solid line represents the shape with minimum drag for a given diameter and is known as a *von Kármán ogive*; the dashed line represents the shape with minimum drag for a given volume.

more accurately, it is possible to say whether the bullet is intended to travel at subsonic or supersonic speeds. Thus, a low-speed (short-range) match quality bullet has a flat base, short shank or body, and a moderate-length nose, whereas a high-speed (long-range) bullet will be boat-tailed with a long shank and a long, pointed nose. Higher speeds are more achievable with more pointed, cone-shaped noses, as we saw in the preceding chapter. (This is true for artillery shells as well as for bullets.

Supersonic rounds often have a conical nose, whereas subsonic howitzer rounds are stubbier.) Benchrest target shooters know a lot about bullet shape, as they often make their own ammunition.[3] Their obsession is long-range accuracy, and so the aerodynamic drag of a bullet is paramount because the shooter would like to minimize bullet drop, drift (to which we soon turn—that's a pun), and other effects that cause spreading of a group of shots from their custom rifles.

The shape chosen for a projectile nose cone also depends upon other factors besides aerodynamic efficiency. For example, the nose cone shape of a bullet has an impact (literally) upon the bullet's effectiveness when hitting a target. Such terminal ballistic effects may be unimportant for target shooters, but they certainly do matter to hunters and to combat soldiers. I will take you on a tour through the complicated territory of terminal ballistics in chapter 7.

A SHORT INTRODUCTION TO LONG RANGE

The complications of long-range ballistics were unknown before World War I because ballistic trajectories were not long enough for these complications to present themselves. You can see from the graph in figure 5.2 how naval gunnery ranges increased during the twentieth century; this is the period during which long-range effects (the most important of which are explained in the next sections) became apparent, and then understood.

Many of the long-range effects of aerodynamics arise because of the shape and size of the projectile. Earlier, and in the technical notes, I have treated the projectile—be it a bullet, a shell, or even an arrow—as if it were a point object. In fact, of course, these projectiles are spatially extended, and that considerably complicates their mathematical description. We have seen how shape and size influence aerodynamic drag, but these factors are just a part of the problem. From figure 5.3 you can see how the number of variables (the *degrees of freedom*, to a mathematician or a physicist) increases when we look at external ballistics in enough detail to require specification of projectile dimensions. A point-like projectile is specified by three variables, say (x,y,z), which tell us its precise location in

3. Target shooters favor these *handloads* because they help achieve consistency (by minimizing irregularities that ruin accuracy), are more aerodynamic, and are economical.

three dimensions. The rates of change of these position variables tell us the projectile velocity. A point-like projectile thus has three degrees of freedom (DoF). Neglecting sideways movement means we can reduce this to two DoF, as in technical note 17, for example, where only x and z variables were described. The spatially extended projectile of figure 5.3 requires more variables because, unlike its point-like cousin, the position of an extended projectile is specified by rotation as well as by translation (linear movement) along each of the three axes. As shown in figure 5.3, we call the linear movements along the (x,y,z) axes *surge*, *sway*, and *heave* and the rotations about these axes *roll*, *pitch*, and *yaw*. Thus, a spatially extended projectile is described by six DoF.

Ballisticians tend not to use the translational terms *surge*, *sway*, and *heave*. Instead of surge and heave—motion defining a trajectory that lies in a vertical plane—we refer directly to the forces that act upon a projectile in this plane: gravity, lift, and drag. Instead of sway we refer to drift, the force that pushes our projectile in a sideways direction, away from the plane. For the rotational degrees of freedom, however, we stick with the mathematicians and physicists by referring to two of the rotations as yaw and pitch. We draw the line at roll, preferring *twist*.

So what is the big deal about bullets and shells being spatially extended? Haven't we already accounted for this in our examination of aerodynamic drag effects? Indeed we have not: there is a lot more juice to be squeezed

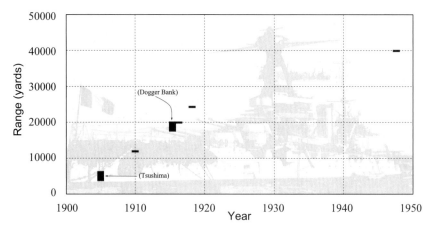

Figure 5.2. Increasing naval gunnery ranges, 1900–1950. Broad bars correspond to varying ranges during the naval engagements shown. Thin bars are from test firings. Data from U.S. Navy, Department of Ordnance and Gunnery (1955).

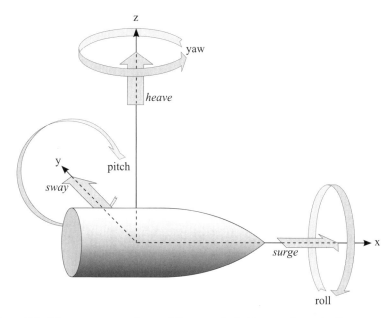

Figure 5.3. The movement of a spatially extended object such as a bullet or artillery shell is described by six degrees of freedom (DoF): linear motion along each of the three directions (*x*,*y*,*z*) and rotational motion about each of these axes. The names given to these DoF are shown.

out of this particular orange. We will see that rifling causes a speeding bullet to *precess*—to wobble slowly. The wobbles may settle down (for a stable trajectory, they will do so) but with the bullet flying through the air pointing in a different direction than its line of flight. The difference is small, but significant for long flights. Because of its shape, a bullet that is not pointing along its direction of motion will present an asymmetric profile to the wind, and this results in sideways motion—drift.

To gain an understanding of how these strange phenomena arise, we must digress briefly to consider spin stabilization and the *gyroscopic effect*. The physics may make your head spin too, but I will keep the discussion brief and avoid the math (because it gets pretty esoteric).

SPIN STABILIZATION AND THE GYROSCOPIC EFFECT

Gyroscopes, toy tops, and the like exhibit behavior that is counterintuitive, and so it is worthwhile to see how they work. To follow this discussion, you will need to appreciate the concept of vectors. It will help if you have a basic understanding of torque and angular momentum; a brief summary of these physical ideas will suffice here.

Torque is the rotational equivalent of force, and angular momentum is the rotational equivalent of linear momentum. A bowling ball plowing through a row of pins exhibits linear momentum; a fast-spinning and very heavy turbine blade exhibits angular momentum—and both are very hard to stop. Force acts to increase or decrease linear momentum; torque acts to increase or decrease angular momentum. So much for the preamble; let's see how these ideas apply to gyroscopes.

Figure 5.4a shows a gyroscope (on a table top) with its central axis leaning over at an angle. If the flywheel is not spinning, the gyroscope will do what everything else that leans over at an angle would do: it falls over. The familiar yet very weird property of spinning gyros is that they do not fall over; the effect of gravity is to cause the gyro to "fall sideways"—to precess about a vertical axis. Precession means that the top of the gyro axis moves slowly around a circle centered above the gyro base, which does not move. In other words, the action of gravity causes a spinning gyro to move perpendicular to the direction of the gravitational force. This is very strange, and the explanation lies with the gyro's angular momentum.

Angular momentum is a vector quantity, as is force. The direction we associate with angular momentum is given by the *right-hand rule*. As shown in figure 5.4a, if the fingers of your right hand curl in the direction of gyro spin, then your upturned thumb defines the angular momentum direction. In this illustration we have a counterclockwise-spinning gyro (when viewed from above), and so its angular momentum points upward, in the direction of the gyro axis. This angular momentum vector is shown in figure 5.4b. Faster spin means higher angular momentum, which we represent by a longer angular momentum vector.

Let us say that the amount of angular momentum possessed by our spinning gyro is J_1. Gravity pulls the gyro down, as shown by the downward-pointing arrow of figure 5.4a (representing the gravitational force vector acting at the gyro center of gravity). The gyro does not fall through the table (at least none that I have seen ever did so) because the table pushes back, in accordance with Sir Isaac Newton's famous third law. This reaction force is shown as the upward-pointing arrow in figure 5.4a, acting at the point of contact between table and gyro axis. Note that the two forces have the same magnitude and opposite directions, as they must, but that they *do not act at the same point*. This means that the two forces generate a torque—they cause the gyro to twist. According to the right-hand rule, this

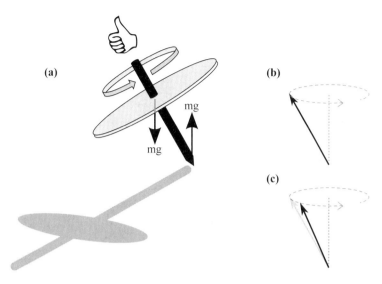

Figure 5.4. The gyroscopic effect. (a) A gyroscope spinning counterclockwise, as seen from above, possesses angular momentum pointing upward, along its axis, according to the right-hand rule. A gravitational torque adds an additional component of angular momentum, but in a different direction: in this case the added component is pointing out of the page. (b) The intrinsic angular momentum vector, J_1, of the gyro flywheel. (c) The angular momentum vector, J_2, of the flywheel plus the extra kick provided by gravitational torque. As time progresses, the gyro axis precesses about a vertical line. This is the gyroscopic effect: movement is perpendicular to the direction of the applied force (in this case gravity).

torque induces a change in the direction of the gyro angular momentum in which direction? Check for yourself: in figure 5.4a the gyro angular momentum is moved out of the page.

Stare at figure 5.4 long enough and you should be able to convince yourself that the new angular momentum of the gyro—call it J_2—is displaced from J_1 as shown in figure 5.4c. Gravity still acts upon the displaced gyro, however, and this time the angular momentum vector (call it J_3 though it is not shown in the figure) is moved further around the circle of figure 5.4c. Thus, the angular momentum vector precesses about a vertical axis. The gyro axis points in the same direction as the angular momentum vector, and so the gyro itself precesses about a vertical axis, under the influence of gravity, instead of falling down.

This explanation skates over a few details but is basically right. A detailed mathematical analysis provides quantitative predictions—borne out

by experimental observations—of the precession rate and of other complications that arise when the gyro flywheel slows down enough so that
the angular momentum vector is small.[4] (The precession becomes unstable, and the gyro falls over).

You may be ahead of me here, as I turn to the ballistic application of
gyroscope phenomena. Rifling causes a bullet to spin fast; it flies through
the air like a gyro on its side. In the next section we see how the gyroscopic
effect—the axis of rotation moving perpendicular to the applied force—
influences bullet trajectories.

LONG-RANGE AERODYNAMIC EFFECTS
Stability

The shape of all modern bullets (and of historical ones after the Minié
ball) and all artillery shells is determined by internal ballistics and terminal ballistics considerations, as well as by aerodynamics, as I said earlier.
Because of the shape of these projectiles, their CG is behind their CP and
they are aerodynamically unstable. They must be stabilized gyroscopically,
or they would not fly straight: they would tumble erratically, considerably
reducing both range and accuracy. But this stabilization comes at the price
of a greatly complicated bullet trajectory. Small forces arise, as we will see,
that deflect the bullet trajectory from the simple two-dimensional trajectory determined by gravity, drag, and lift. These deviations become apparent over long trajectories, and a shooter or gunner must allow for them.

How does gyroscopic stabilization work? We have seen that a gyroscope
does not fall over when its flywheel spins fast; the effect of gravity is to
push the gyro's CG sideways. The gyro top then follows a circular precession, centered on a point directly above the stationary base. A bullet with
spin possesses a large amount of angular momentum, as does the gyro, but

4. The definitive work on the physics of gyroscopes, the monumental *About the
Theory of the Gyroscope*, was written between 1898 and 1914 by two German
mathematical physicists, Felix Klein and Arnold Sommerfeld. Less eminent physics
students, struggling with the mathematics of gyros (which involve vector cross
products and the right-hand rule), have been seen putting down their pens during
examinations and manipulating their right hands in front of their eyes, in a manner
that must appear bizarre to observers unfamiliar with vector dynamics. Readers
who are seriously interested in the mathematics and physics of gyroscopes would
be better advised to turn to a modern physics textbook on the subject, such as
Kibble and Berkshire (1996).

from the right-hand rule you can see that the direction of the bullet's angular momentum vector is forward (i.e., approximately horizontal, for a flat bullet trajectory). All American rifling is right-handed, imparting a right-handed, or clockwise, twist to the projectile. (For a left-handed twist, the projectile's angular momentum vector would point backward.) It is the large value of angular momentum (directed forward or backward—it doesn't matter which—more or less along the direction of flight) that provides flight stabilization.

In theory, a bullet emerges from the barrel muzzle pointing exactly along the direction it is moving. In other words, the bullet's velocity vector and angular momentum vector are pointing in exactly the same direction. But even if this is the case initially,[5] it does not last. The bullet's velocity vector will tip downward as the bullet arcs through the air. So, in practice there is usually a small angle (a degree or so) between the direction in which the bullet points (read "angular momentum vector") and the direction in which it moves (read "velocity vector"). This matters because unless the two vectors are aligned, the wind force acting on the front of the bullet is asymmetric. The wind force is composed of two components, recall: drag and lift. These forces are shown in figure 5.5a.

You can see from figure 5.5a that because the bullet nose is above the velocity vector, there will be a nonzero lift force that causes the bullet to deviate from its original trajectory. Also, the drag force will cause the bullet to *overturn*—to tumble (fig. 5.5b). Neither of these occurs if the nose points exactly along the velocity vector, because the bullet is then symmetric about its longitudinal axis, but tumbling and lift occur for the usual case when the angle *a* (the yaw angle) is not zero. Tumbling does not occur, and most of the lift is avoided if the bullet is spinning. The drag force does for a spinning bullet what gravity does for a gyroscope; it causes the bullet's rotation axis to precess about the line of force. So, the nose of the bullet shown in figure 5.5a will rotate about the velocity vector with angular radius *a*; it will not tumble. (Applying what we know about gyroscopic effect, you can see that our bullet will precess counterclockwise, as seen from in front, fig. 5.5c.)

5. It is usually not the case; a bullet emerges pointing in a slightly different direction than that in which it moves, for several reasons. The violence of the muzzle blast can throw a bullet a little to one side or cause it to tilt. Imperfections in manufacture, exacerbated by rapid spinning, may induce secondary rotations, and so forth.

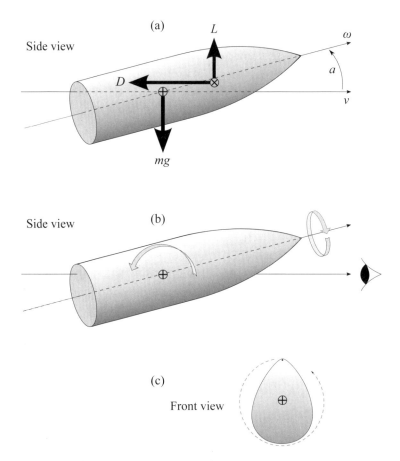

Figure 5.5. (a) Lift, drag, and gravitational forces acting upon a bullet with yaw angle *a*. Gravity acts at the CG (+) whereas lift and drag act at the CP (×). Bullet angular momentum is the direction of the vector *ω*. (b) The overturning moment, or torque, which adds a component of angular momentum (pointing out of the page) to the bullet's angular momentum arising from twist. From the point of view of the eye pictured, the bullet precesses as shown in (c).

If enough twist is applied to a bullet so that it precesses instead of tumbles, the bullet is said to be gyroscopically stable. As the velocity direction changes—it drops as the bullet is drawn down by gravity—the bullet's longitudinal axis (and angular momentum vector) is drawn down with it, maintaining the precession.[6] A well-designed bullet will in fact

6. An *overstabilized* bullet has too much twist, with the result that the direction of angular momentum does not change throughout the trajectory. In a high-angle

precess with reducing angle a, so that the nose spirals toward the velocity vector. When this occurs, as is usually the case, then the bullet is said to be *dynamically stable*, as well as gyroscopically stable.

Drift

In practice, bullet orientation does not quite spiral down to zero yaw angle ($a = 0$); it settles down to a small but nonzero angle called, rather poetically, the *yaw of repose*. Because of the overturning torque (fig. 5.5a) this yaw angle is such that the bullet nose is to the left of the velocity direction as seen from the front (to the right, as seen from behind). Consequently, the lift force acts to push the bullet—actually push the bullet, not just cause it to rotate—so that the shooter notices the bullet drift to the right. This drift has nothing to do with crosswind—it occurs in still air—but is a consequence purely of the aerodynamics of a spinning projectile.[7]

Data from an early French rifled musket of 1842 (fig. 5.6) shows how drift increases with distance from the muzzle. This data was reported by Benton (1862) at a time when twist was not understood theoretically. He reported further U.S. tests with a rifled musket in which the mean drift distance (of 40 shots) at a range of 1,100 yards was about 18 feet with not a single shot deviating to the left. Benton was able to assure his West Point readers that the effect was not due to recoil, as some people had once thought, because the French data showed that drift increased with distance and therefore was an aerodynamic effect.

I have now described one of the main aerodynamic complications that arises with long-range trajectories. Twist leads to precession, which leads to drift, due to lift. As the bullet precesses, the bullet trajectory corkscrews through the air. The precession settles down to a stable (constant) yaw of repose, and so the lift force produces a small but significant drift, to the

———

shot such a bullet may land base first. This is not good: a lot of speed is lost in the final stages of such a trajectory because of the larger bullet cross section presented to the air, and the penetrating power of the bullet is reduced. Handgun bullets are usually overstabilized, but it doesn't matter because handgun ranges are so short. A bullet which tips over as its velocity vector tips over, so that its nose lies on the line of flight, is said to have a *tractable* flight—and this is the ideal case. In practice, the precession is not always circular as with the gyroscope, but describes a more complicated revolution.

7. Of course, a crosswind will also push the bullet to one side, but that is an additional and quite separate consideration.

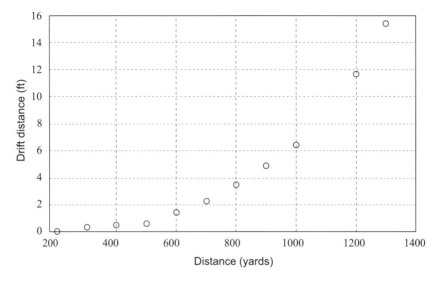

Figure 5.6. Drift distance (feet) vs. distance from muzzle (yards) for a nineteenth-century French rifled musket. Data from Benton (1862).

right for a right-handed twist. The amount of drift depends upon the bullet's size, shape, weight, and twist.

The Magnus Effect

The Magnus effect results from another aerodynamic force that makes projectiles do unexpected things. A golf ball with backspin gains lift, a baseball curves sideways, and a spinning bullet or artillery shell moves in various directions due to the *Magnus force*. As with drift, the Magnus force acts mostly in a direction that is perpendicular to the velocity vector, and so a projectile is pushed up or down or sideways. The force is small compared with that which gives rise to drift, but on long trajectories it can be significant. To see how the Magnus force works, consider figure 5.7.

The side view shows a bullet with its longitudinal axis at angle *a* to the velocity vector. A precessing bullet in flight will be like this—with its nose above its CG—more often than it will be found in other orientations because the velocity vector is always turning downward as the bullet falls under the influence of gravity. From the bullet's point of view, air rushes past it from right to left, for the case shown in figure 5.7. The air that passes close to the bullet is dragged around as the bullet spins, as a result of skin friction. Viewed from behind the bullet, the air streamlines are as illustrated. (In the drawing it appears as if the air is rising up from the ground. In fact, the biggest component of air velocity is out of the page, in this back-view picture, with only a residual component seeming to move up around the bullet, due to the small angle *a*.) The air moving up the left side passes the bullet surface with lower relative speed than the air moving up the right side because of bullet spin. Slower air moving over a surface sticks to it more strongly than fast air. This is a well-known aerodynamic phenomenon, known as the *Coanda effect*. The effect is that when the air detaches from the surface, it does so more quickly on the right side than on the left, so that the streamlines above the bullet are deflected to the right,

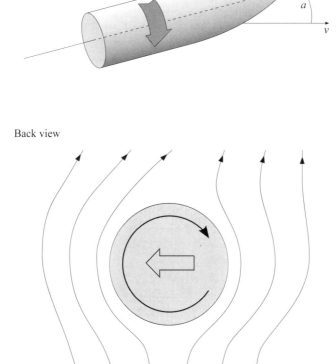

Figure 5.7. The Magnus effect arises from air flow over a spinning surface. The side view shows a bullet with yaw angle *a* and right-hand twist. The back view shows that air flowing past the bullet is deflected to the right. Consequently the bullet is pushed left.

as shown. Newton's third law then requires that the bullet move to the left, to counter this transfer of momentum. This is the Magnus effect.[8]

For the case shown in figure 5.7, the Magnus force acts in the direction opposite to the drift force (i.e., the lift force that arises due to gyroscopic precession). But because it is smaller in magnitude, it only slightly reduces the drift effect. Had I shown, in figures 5.5 and 5.7, the bullet oriented differently—say, at its angle of repose (the bullet pointing to the right of the velocity vector, and a little above it)—then the Magnus force would be directed almost vertically up and so would be perpendicular to the drift force.

There are at least a half dozen other aerodynamic forces and torques that act upon a spinning bullet, but their effect is even smaller than the Magnus effect, and so I merely list them:[9]

- Spin damping moment. Due to drag, this force slows down the rate of spin.
- Magnus moment. If the Magnus force vector does not pass through the bullet's CG, it generates a small moment.
- Pitching force. This force opposes the change in direction of the projectile's longitudinal axis.

8. I have here provided a "momentum transfer" explanation of the Magnus force. Some explanations make use of Bernoulli's law, which says that higher speeds generate lower pressures, pulling the bullet to the left. This explanation is wrong. In fact, the air pressure may be reduced on the left side of the bullet, but this is what causes higher air speeds (instead of higher speeds causing lower pressures). Much of the confusion that exists in popular literature is due to misapplication of Bernoulli's law. For a fuller explanation of the Magnus effect, and of the popular misconceptions concerning aerodynamic forces in general, see the appendix to my book of the physics of sailing (Denny 2009). For detailed explanations of the aerodynamic forces that act upon projectiles, see Canada DND (1992). A full mathematical treatment is provided in the appendix to the U.S. Department of Defense handbook on fire control systems (1996).

9. If you want to know more about these forces and torques, consult the appendix to U.S. Department of Defense (1996). See also Nennstiel (1996) for a detailed account specifically of the external ballistics of handguns and rifles. Nennstiel quotes experimental confirmation of the theoretical developments and concludes that while short- and medium-range trajectories are pretty well understood, there is as yet no fully reliable theory of long-range trajectories. Farrar, Leeming, and Moss (1999) is a good textbook account of all the details. Both these references are heavy on math.

- Damping moment. This is associated with the pitching force.
- Magnus cross force and moment. These arise from the transverse angular velocity of the projectile longitudinal axis.

Nutation

Precession is sometimes called "slow mode oscillation" by ballisticians. There is in general also a "fast mode oscillation" called *nutation* by physicists. This extra mode is superimposed upon the smooth circular precession of figure 5.5c (in reality a less regular circumnavigation of the CG), appearing as a sinusoidal or looping motion. So the front view of the nose cone movement of a bullet undergoing precession and nutation would be like a wavy line or curlicue bent around into a circle. Nutation arises because the spinning bullet does not in general start out with its nose right on the precession circle; it has to get there, because of the gyroscopic effect, but it overshoots due to its own inertia and subsequently oscillates as it attempts to reach the circle. (Think of a tetherball: struck at the correct angle it describes a smooth circle about the pole, but struck at a different angle it has a more erratic trajectory.) I don't consider the nutation motion as important as precession because the influence of nutation on bullet trajectory is less important: the fast oscillations average to nearly zero over one precession cycle.

LONG-RANGE EARTH CURVATURE EFFECTS

At ranges longer than any rifle achieves, we encounter phenomena that influence a projectile trajectory. These phenomena, outlined below, become important for the accuracy of long-range artillery (an artillery crew with their gun at location A wants to hit a target, say a building, at location B). They arise because of our planet's finite size and its rotation.

Effects of Earth's Radius

We examined ballistics on a flat earth with no atmosphere in technical note 15. Determining range on a curved earth is more complicated, even with the simplification of no atmosphere. The analysis, presented in technical note 19, shows that, for the same muzzle speed and elevation angle, the range of a gun on a curved earth is longer than that computed for a flat earth. For example, on a flat earth, a muzzle speed of 3,000 ft/s leads to a maximum range of 93,750 yards, neglecting air resistance. On a curved earth the same gun has a maximum range that is 630 yards farther. Of course, both ranges (and the range difference) will be reduced by the effects of air resistance, though not so much as you may think, as we will soon see. The point is that our gun, if it is aiming at a specific target such as a building, will need to account for earth curvature or the shot will fall several hundred yards beyond the target.

The reason that finite earth radius has this effect is clear: over long distances the earth "falls away" beneath the artillery shell as it flies though the air, so that the shell is still airborne at the range where, on a flat earth, it would have hit the ground.

Earth Rotation: The Coriolis Effect

The Coriolis effect seems rather strange at first, but the explanation—if not the math analysis—is straightforward. The Coriolis force is what physicists call a *fictitious force* because it arises as a consequence of the observer's movement (acceleration, more specifically). Centrifugal force is another example of a fictitious force: it feels real enough to someone who is moving at speed around a curved path—say, driving a car around a curve—but it is purely a consequence of the movement. Both centrifugal and Coriolis forces apply to objects on the earth's surface, because the earth's surface is rotating. Because of centrifugal force, gravity appears to be a little less strong at the equator than at the poles: a person on the equator is thrown outward, and this partially cancels the pull of gravity. In addition, centrifugal force produces an equatorial bulge (the earth is not spherical—it is slightly squashed at the poles); thus, the equatorial surface is farther from the center of the earth than is the polar surface, and gravity is weaker.

Centrifugal force, while familiar to us, is not important in the context of ballistics. The much less familiar Coriolis force is significant, however, for very long-range artillery shells or unguided (i.e., ballistic) missile trajectories. We can see how it works from figure 5.8, which shows a bug on a turntable, with a bubble floating in the air just above the turntable. Assume that the bubble moves in a straight line at constant speed and that the turntable rotates at constant angular speed. From the point of view of the bug in figure 5.8a, the bubble moves along the curved path shown in figure 5.8b. (To help fix the idea in your mind, I provide a second example: from the point of view of the bug in figure 5.8c, the bubble moves along the curved path shown in figure 5.8d.) The dotted lines of figure 5.8a and b are constructions to demonstrate that the relative positions of bug and

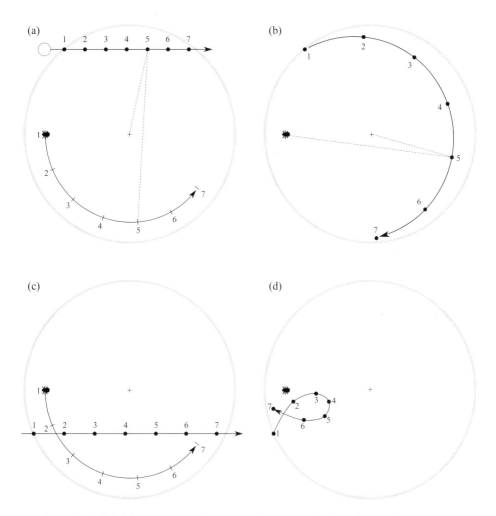

Figure 5.8. (a) A bug on a rotating turntable and a bubble floating past freely as viewed by an observer who is standing still. Numbers indicate bug and bubble positions at different times. (b) The bug's-eye view: to the bug the bubble appears to have a curved trajectory, which means that a force must be deflecting it. This is the Coriolis force. The construction (dashed lines) helps to show that the angle and distance from bug to bubble (here shown for position 5) is the same in both frames, which confirms that the two different views are of the same thing. (c) Another example: the bubble starts from a different position. To the bug the bubble appears to move as shown in (d).

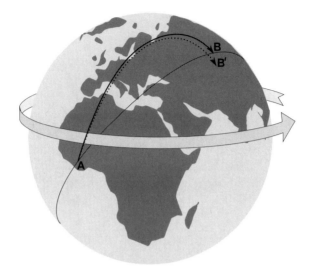

Figure 5.9. The Coriolis force causes a projectile in the Northern Hemisphere to veer to the right of its intended target. The intended trajectory is shown as the solid line, while the actual trajectory is the dashed line.

bubble are the same in both cases—here shown when both are at position 5 of their individual trajectories. You, standing on the floor and not on the turntable, see that the bubble moves at constant speed and in a straight line (fig. 5.8a and c), and so you consider that the bubble is not acted upon by any force. The bug, on the rotating turntable, sees the bubble move along a curved path (fig. 5.8b and d) and so considers that the bubble must be acted upon by a force (we have an intelligent bug here). This is the Coriolis force. It is fictitious because it depends upon your point of view (your *reference frame*, in the language of physicists and mathematicians).

Now consider the rotating earth (fig. 5.9). Viewed from above the North Pole, the situation of people on the surface is very like that of the bug on the turntable. If we fire an artillery shell from A to B, as shown, it appears to veer away from the expected trajectory as a result of the earth's rotation, just as the bubble seemed to follow a curved trajectory as seen by the bug. There is an extra complication for the rotating earth, however, in that the earth is approximately spherical whereas the turntable was flat: we are now dealing with a three-dimensional Coriolis force instead of a two-dimensional force. The idea is the same, nevertheless.

The detailed math analysis is complicated, but it allows us to make precise predictions of the effect of Coriolis force upon ballistic trajectories. The general rule is that trajectories veer to the right in the Northern Hemisphere and to the left in the Southern Hemisphere. For example, in figure 5.9 a gun crew at A aim their (very) long-range gun at a target B; it will land somewhere to the right of where they expect (say at B′) if they do not account for Coriolis. Here are a couple of specific examples. Say a rifle fires a NATO 5.56 mm round due north with a muzzle speed of 970 ms⁻¹ (3,180 ft/s) at an elevation angle of 0.1°. The range (neglecting air resistance) is 335 m (366 yd), and the Coriolis force pushes the bullet 6 mm (0.24 in) to the east. This is negligible: we can forget Coriolis for short ranges. My second example is a long-range artillery gun with muzzle speed of 1,000 ms⁻¹ (3,280 ft/s) aimed north with an elevation of 45° for maximum range. Neglecting air resistance the distance traveled is 102 km (64.4 miles), and the shell lands 504 m (551 yd) east of its intended target. This distance is significant.

To illustrate the complexity of Coriolis forces, consider now what happens when the same artillery piece is fired eastward. This time the shell lands 1,069 m (1,169 yd) south and 378 m (413 yd) short of its intended target: the effect of Coriolis depends upon direction aimed.

Atmospheric Attenuation

Gravity pulls the earth's atmosphere down, so that the density of air changes with height above the surface. Atmospheric density is about 1.23 oz/ft³ (or 1.23 kg m⁻³) at the surface and drops by half for every 3.2-mile (5.1-km) increase in altitude. At the rarified altitudes of the ballistic trajectories of long-distance artillery shells, aerodynamic drag is greatly reduced; consequently, it is only at the beginning and end of such trajectories that drag is a significant factor. I can best illustrate the importance of atmospheric attenuation with altitude by telling you something about a famous long-range artillery weapon, the Paris Gun used by the Germans in World War I to shell Paris. Its range was so long that the gunners needed to take into account all of the effects discussed in this section: drift, earth radius, the Coriolis effect, and atmospheric attenuation.

THE PARIS GUN

In the last year of World War I the Germans decided that it would be a good idea to shell Paris. This would, German leaders felt, be a morale

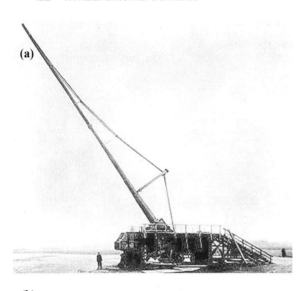

Figure 5.10. The Paris Gun, used against Paris in World War I. (a) One of the new surviving photos shows the bracing required to keep the 112-foot-long barrel straight. Adapted from a Wikipedia image. (b) A shell cross-section. Note the long hollow nose cone. Image from Wikipedia.

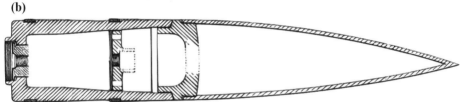

booster for the German people. Trouble was, the German front line was almost 60 miles from Paris. This receded to nearly 70 miles after the ill-fated French Nivelle offensive of 1917, and so Erich Ludendorff, commander-in-chief of the German army and de facto government leader, telegraphed a Dr. von Eberhardt at the Krupp armament facility and said, "Make that gun shoot 75 miles." Eberhardt had originally proposed a gun with a range of 60 miles and was busy developing it, so this telegram came as something of a jolt. The weapon was made, however; and, manned by a crew of 80 from the German navy, it began firing upon Paris early one morning in March 1918 from a range of 75 miles (120 km).[10]

The Kaiser Wilhelm Geschütz, to give this enigmatic gun its German name, was a 21 cm caliber rail-mounted gun weighing 125 tons (fig. 5.10). Its barrel length was 34 m (112 ft). The 120 kg shell was propelled out of the muzzle by a 180-kg charge, at around 1,600 ms^{-1} (i.e., a 264 lb shell, a

10. A history of the Paris Gun and its deployment is provided by Marshall (1987).

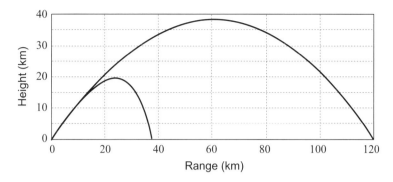

Figure 5.11. Height vs. distance for the Paris Gun, with and without atmospheric attenuation. Attenuation meant that most of the trajectory took place in a near vacuum, so drag was reduced and range was greatly increased.

396-lb charge, at 1 mile/s). This monster (the second largest rifled weapon to be used in warfare, behind a German World War II howitzer called the Schwerer Gustav) sent its shell to the edge of space in a 3-minute flight that ended 75 miles away. Given the size of the gun and of the propulsive charge, the explosive charge delivered to the target was feeble: 7 kg or perhaps 15 kg (the sources are not consistent). Between 320 and 367 shells landed on Paris, at a rate of 20 shells per day, killing about 300 people and wounding over 600. Eventually the gun position was overrun by the Allies (it was only 7 miles behind the front line), though the gun itself was never found.[11]

The maximum height of the shell trajectory has been calculated to have been about 40 km (130,000 ft). Most of the trajectory was in the strato-sphere, and so the air resistance was negligible. I have simulated this trajectory on a computer (fig. 5.11), and, indeed, that is the height obtained. The simulation also reproduces the 3-minute flight duration. The near-parabolic shape of the true trajectory indicates that most of the flight takes place in a vacuum. From the figure we can easily see the significance of atmospheric attenuation: with no attenuation (i.e., assuming that drag throughout the flight is the same as occurs near the surface) the range of this gun is reduced to less than 40 km (25 miles).

The elevation of the Paris Gun was fixed at 50°. The range was altered by adjusting the propellant charge. In fact, for these long-range trajecto-

11. In fact, there were several guns. The wear and tear on the barrels was considerable, and they needed to be rebored after 65 shots.

ries, the optimum elevation angle for maximum range is about 52°, as simulation reveals, so the gunners got it about right.

The long shell was stabilized gyroscopically, though this was easier in the rarified atmosphere. One problem in theory is that the stabilization required would change with altitude, but it seems not to have arisen in practice. A technical analysis of the weapon was carried out immediately after the war by a British major, J. Maitland-Addison. He concluded that the effect of the earth's curvature must have been included in range calculations (it would add half a mile to the range) and allowance must have been made for the Coriolis force (which, he ascertained, would have thrown the shell 400 yards to the right and extended the range by 700 yards). Drift would have been less than for low-altitude trajectories.[12]

Despite the careful calculations, it seems that the Paris Gun was not very accurate—hardly surprising, given the extreme range. It could be expected to hit a target the size of a city, but the distribution of shell explosions seems to have been random, hitting churches and other civilian targets. Random atmospheric effects over the long flight, as well as variations in propulsive charge and shell construction, plus barrel wear, would all have contributed to this randomness.

The aerodynamic performance of a ballistic projectile is influenced by its nose cone shape; the optimum shape depends upon projectile speed. Stabilization of projectiles is achieved by rifling and depends upon the gyroscopic effect. Spinning projectiles lead to two aerodynamic forces that are significant for long-range gunnery: drift and the Magnus effect. (Other, smaller forces and torques also arise.) Three other phenomena influence long-range trajectories, and these are the result of the shape and size of our planet: the effect of the earth's radius, the Coriolis effect, and atmospheric attenuation. All of these effects contributed to determining which *arrondissement* (district) the projectiles from the Paris Gun would hit.

12. See Maitland-Addison (1918).

6 New Technology, New Ballistics

Weapons technology progressed rapidly in the second half of the nineteenth century, as we have seen. Innovation and technological advances continued throughout the twentieth century, at a somewhat slower pace, but the development of weapons platforms and integrated weapons systems accelerated, and this explosion (pardon the pun) of weapons technology continues today. Bigger guns fire more rapidly to longer distances, delivering more devastation from more mobile platforms. Artillery was horse-drawn at the end of the nineteenth century, was mounted in a tank by the mid-twentieth, and became airborne by the end of the twentieth century. The rate of fire and the effective range of automatic and semi-automatic firearms increased, even as these hand-held weapons became lighter to carry.

All these developments made the problems associated with weapon aiming more acute. Increased ranges, increased target speed and maneuverability, increased weapon platform speed and maneuverability—all made the *fire control* problem more difficult to solve. In the first part of this chapter we will examine some of the solutions that were developed, particularly for artillery and for airborne weapons, to assist the delivery of ballistic munitions to their increasingly elusive targets.

The twentieth century saw the development of effective rockets. Rocketry changed the game, slowly at first but then at an increasing pace. In the second part of this chapter we will see how rockets work and why their ballistics are different from those of gunpowder projectiles. I end the chapter with a summary of "the end of ballistics": the solution to the increasingly complex ballistics problems at longer and longer ranges that we saw in chapter 5. This solution takes the form of guided munitions, which are far more accurate than ballistic missiles over thousands of miles. The introduction of remote sensing systems combined with computer-controlled navigation and guidance systems means that the longest-range missiles have to solve a different set of problems from those faced by the crew of the Paris Gun, for example. These problems fall outside the field of ballistics, strictly speaking, and so here I will only summarize the key features of guided weapons.

FIRE CONTROL

The key feature of weapons construction that made fire control both possible and necessary is consistency. If the barrels of two nominally identical guns are different or if the rounds they fire vary in size or charge or shape because of significant manufacturing imperfections, it is impossible to provide accurate methods of fire control. Consistency from one weapon to the next permits the development of aiming devices that can be applied to all the weapons of a given type. Of course, each rifle or artillery piece still has to be individually aimed, but a dedicated aiming device can be adjusted to cover the small variations between weapons (due to barrel wear, for example). This consistency of manufacture, through precision engineering, first appeared at the end of the nineteenth century, and so the same period saw the introduction of effective gun sights, range finders, gunner's quadrants, and other equipment (such as fuze settings) that led to increasingly accurate and effective munitions delivery.

In the eighteenth century and earlier, few firearms were provided with sights. There was no point (another pun) because, as we saw in the first part of this book, the internal ballistics were such as to render a firearm weapon intrinsically inaccurate. Muskets were aimed at blocks of infantry or cavalry, not at individuals. In the nineteenth century rifles and—later on—artillery were provided with aiming sights. At first these were simply front and rear sights that enabled the shooter or gunner to align the barrel of his weapon with the target. Later still the rear sight became adjustable, so that the weapon could be adjusted in elevation for range, or aimed off-target in azimuth (i.e., horizontally) to allow for the effects of twist or of crosswind. By the end of the nineteenth century, artillery weapons were provided with sighting telescopes with adjustable mountings. Azimuth and elevation sights were operated independently; the gun crew would include a trainer, who aimed the weapon in azimuth, and a pointer, who was responsible for gun elevation.

In the twentieth century, range-finding equipment was introduced—increasingly necessary as ranges got longer. Gunners' quadrants permitted the setting of barrel elevation with unprecedented accuracy, optical range finders first appeared, and triangulation techniques were developed, such as the "sound-and-flash" method of ranging enemy artillery, examined in the next section.[1]

Gyro gun sights helped to keep barrels level and on target. Originally these were applied to naval weapons; in the Age of Sail, gun crews would wait for the middle of the roll of their ship before firing cannon, but by the twentieth century naval guns could be kept level in a rolling sea and so could be fired at any time. By World War II this gyro technology was being applied to tanks in the deserts of North Africa and to the fighter planes that played such a significant part in the war.

Before World War II, weapons platforms had been stationary, and their targets were either stationary or moving slowly. A rolling sea, rough terrain for a tank, or a maneuvering airplane presented new problems that were greatly ameliorated by gyro gun sights. We saw in chapter 5 how the gyroscopic effect stabilizes spinning bullets; the same effect can be used to stabilize weapons-aiming platforms. A fast-spinning gyroscope maintains a fixed orientation in space and resists changes in orientation, say, as a ship or airplane rolls. By World War II gyro gun sights fitted onto fighter planes enabled them to dogfight more effectively. In figure 6.1 you can see an example: a gyro gun sight is fitted in a Spitfire cockpit.[2]

Gyro sights could compensate automatically for both target lead and bullet drop. The pilot would place his airborne target in the gun sight cross-hairs, but his weapons (machine guns or light cannons)[3] would point in a different direction, to allow for target movement and for the ballistic trajectory during the time it took the bullets to traverse the distance to the target. This is getting quite sophisticated: target speed, heading, and range must be known for such aim-ahead gun sights to work. We are entering the

1. Sound-and-flash ranging and other early techniques are summarized in U.S. Department of Defense (1996).

2. The gyro gun sight shown in figure 6.1 was manufactured in Edinburgh, Scotland, from 1943 on at the facility where, 40 years later, your author would begin his career in aerospace.

3. Fighter planes of this period were formidably armed. A Spitfire in the Battle of Britain was sent aloft with eight 30-caliber machine guns, four in each wing; a P-51 Mustang later in the war would have either four or six 50-caliber machine guns. As a physicist I wonder about the recoil effects of all this firepower on airframe speed.

Figure 6.1. Spitfire cockpit with a Ferranti Mk IIc gyro gun sight (box at top of instrument panel, with angled glass viewer). Image from Wikipedia.

age of integrated weapons systems, with electronic technology (crude analog devices during World War II) aiding the pilot and relieving him of some of the burdens of weapon-aiming.

"Thunder and Lightning"

One early range-finding technique, specifically to locate an enemy artillery piece, is the "sound-and-flash" method of triangulation. Let us say that you command a gun crew that is part of a battery of six guns. You observe through binoculars the flash of a distant enemy gun and time the interval from the flash to the sound. For an isolated enemy gun this "thunder and lightning" approach might provide a rough estimate of range. Your results are better if all the gun crews in your battery work together to time the flash, as illustrated in figure 6.2a. Here, your gun and the five others of your battery are strung out along a line, say 200 yards apart. An enemy gun position is located about 1,500 yards away. Each gun crew records the time of the sound of a gun report from the enemy; the sound from a single blast will arrive at your six guns at different times, because of the different distances. Knowing the geometry, you can work out the distance to the enemy position. Estimates based upon the measurements by the individual crews are averaged, thus reducing the overall error and providing a better estimate of enemy position. The example of figure 6.2b shows the results of a computer simulation of this method; you can see that the averaged estimates of enemy position are less spread out than are individual estimates.

A modern version of this technique replaces the gun crews with an array of microphones. The time of arrival of a gun blast from an enemy position is recorded electronically. Time differences, and the known geometry of the microphone array, permit estimation of both the range and the bearing of the enemy. It is not necessary to observe the flash of the enemy muzzle, so this method will work if the enemy position is hidden from view. The modern approach is predicated upon very accurate electronic timing and computing power that can quickly turn the recorded times into position estimates.

The sound-and-flash method worked well enough in an open setting with an isolated enemy. With many guns, however, it would be difficult to know which sound was associated with which flash, and for long-range artillery you may not have line-of-sight. The modern microphone-array method is more robust: it can estimate the positions of several enemy guns that are firing in rapid succession, so long as there is no overlap in sounds from different guns arriving at the array.

Bombs Away

In World War I it was difficult to aim *from* airplanes, as well as to aim *at* them. From the point of view of those on the planes, all their ground targets were moving fast. Bomb-aiming methods early in the war were as crude as the biplanes from which the bombs were launched. The pilot would hold a small bomb out the open cockpit and drop it when he thought it a good idea to do so. Bombs are designed to be aerodynamically stable in that they have fins at the back to help them maintain a streamlined aspect, but these same fins make bombs susceptible to crosswinds. Bombs dropped from altitude were thus not very accurate: if 100 bombs are dropped from an airplane at exactly the same time, they will spread out and not land in the same place. Even if there is no systematic error (say due to crosswind or mistaking air speed) the bombs will spread out randomly around the intended target because of random atmospheric

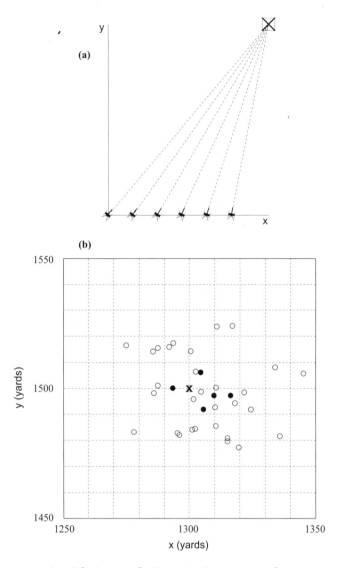

Figure 6.2. Sound-and-flash range finding. (a) A linear array of six guns aim at an enemy gun emplacement (**X**). Each of your gun crews records the time of the sound received from a single enemy blast. From this data and the geometry, it is possible to estimate the enemy range. (b) Results of a sound-and-flash computer simulation, assuming that each of your gun crews can estimate enemy position with an angular error of 10 mils (a mil is $\frac{1}{6400}$ of a circle) and a range error of 1%. The true enemy position is marked **X**. Five estimates from your battery (resulting from five enemy shots) are marked as filled circles; each of these are averages of individual gun crews estimates, marked as open circles. You can see that, by averaging results, the battery comes up with better estimates of enemy position than do individual guns: the spread is reduced.

disturbances. This spread is statistical; the degree of spread (the *standard deviation* of the distribution) depends upon bomb design, airplane altitude, atmospheric disturbances, and other unquantifiable effects, such as the different ways that individual bombs tumble out of the airplane.

To improve accuracy, those early pilots who hand-launched their bombs were obliged to approach their targets from upwind or downwind (thus reducing the effects of crosswind). This might be a problem if the target was able to fire back because they could anticipate the direction of approach of the bombers. By the third year of World War I an effective mechanical bomb-aiming device was in service. The CSBS (course-setting bomb sight, a.k.a. the Wimperis sight after its inventor) of 1917 was capable of drift compensation,[4] eliminating the need to approach a target from a predictable direction. However, it was necessary to approach at constant speed and with straight and level flight, at an altitude below 1,500 feet, so the bomber was still an easy target.

A number of improvements were made between the wars and during World War II, when bombing was upgraded to strategic importance. All these devices were mechanical and based upon optical sighting of the intended target. The 1930s saw the introduction of the preset vector bombsight; its hand settings compensated for bomb ballistics, drift, and aircraft change of heading. The year 1940 ushered in the continuously set vector bombsight, which automatically compensated for aircraft speed and height, plus wind drift, via a mechanical analog computer connected to the aircraft's navigation system. The *tachometric bombsights* of World War II—a family of devices—allowed for measured wind effects. The most famous of this family is the Norden bombsight, which directed most American bombs in both theaters of World War II, the Korean War, and the Vietnam War.[5] This sight enabled a bombardier to measure and compensate for wind drift, and it took over as autopilot during the final approach, releasing the payload of bombs automatically at the right time.

Despite the sophistication of aiming devices, bombing during and after World War II was notoriously inaccurate. German defenses forced American planes, operating in daylight, to higher and higher altitudes, which ad-

4. "Drift" in this chapter refers to crosswind drift and not to the drift resulting from a spinning projectile as discussed in chapter 5. Early bomb-aiming devices are discussed in Baden (1961, chap. 1).

5. The British and Germans also had sophisticated tachometric bomb-aiming devices during the latter stages of World War II.

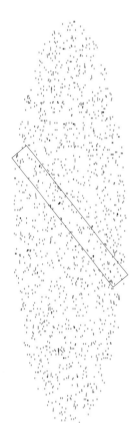

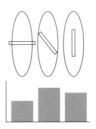

Figure 6.3. An airplane drops hundreds of bomblets; they spread out because of random effects discussed in the text and land in an elongated ellipse caused by airplane speed. The target is a runway (rectangle). Which direction of approach results in the greatest number of hits? You can see from the bar graph that a diagonal approach works best. Perhaps you can consider why the other two directions shown would not be optimum.

versely affected accuracy—and the standard deviation of bomb-aiming errors increases with altitude. (British bombers operated at lower altitude but at night, making it difficult to see the intended target, so they too were very inaccurate.) It has been estimated that only 31% of U.S. Air Force bombs dropped over Germany landed within 1,000 feet of their intended target, despite the boasts of aircrews that they could land a bomb "in a pickle barrel from six miles up."[6] As a consequence, the effectiveness of a bombing raid was often overestimated, particularly as it was difficult to judge damage from the post-op reconnaissance photos. Also, bombed cities from London to Berlin proved remarkably resilient at clearing debris and keeping their infrastructure going.

As an illustration of the complexities involved in effective bombing, consider the problem posed in figure 6.3. Here we have an airplane in

6. See Ross (2002) for an account of U.S. bombing efficacy.

level flight dropping hundreds of bomblets (small bombs) on a linear target—say, an enemy airfield. What is the best angle of approach? Surprisingly, perhaps, there is one.

Dive bombing in World War II was a different story, for two simple geometrical reasons. First, dive bombers released their payloads at low altitude, so the random effects of airflow were minimized. Second, dive bombers approached the target, whether a tank or a ship, at a steep angle, which reduced the spread compared with that obtained for level-flight bombing, as seen in figure 6.4. The pilot dove from altitude (15,000–20,000 feet), pointed his plane at his target, released his single bomb at about 3,000 feet, and then pulled out of the dive in a high-g turn to save himself from hitting the ground or dunking in the sea. (Airbrakes reduced the plane's speed to assist the pullout, and also terrified the enemy below by shrieking like a banshee.) The bombsight had less work to do than for level flight: less drift to compensate for and less curvature of the ballistic trajectory to take into consideration. Another advantage of dive bombing was that targets were often designed to withstand attack from the side, not the top.

Stuka and Dauntless

Perhaps the two best known, and certainly the most important, types of dive bombers were the German Stuka and the American Dauntless. In the first few years of World War II these airplanes defined a type of warfare that later disappeared—or, rather, was made obsolete by guided missiles.[7] The Junkers 87 Stuka was the icon of German Blitzkrieg warfare. Used against land targets such as gun emplacements, this gull-winged two-seater was devastatingly effective during the lightning attacks against Western European nations, as it was earlier in the German invasion of Poland and later during the invasion of the Soviet Union. Because of its weak defenses, the Stuka was less successful when pitted against an enemy with an effective air force, as during the battle of Britain in 1940 and later in the war when the German air force, the *Luftwaffe*, was heavily outnumbered.

The American Douglas Dauntless (fig. 6.5) was a rugged workhorse that was hugely influential in the war against Japan in the Pacific. Sent against Imperial Japanese Navy battleships and aircraft carriers, the Dauntless was

7. We might regard dive bombers as guidance systems that controlled 80% of the bomb's trajectory, leaving only the final 20% as an unguided ballistic trajectory. Japanese kamikaze planes were, in essence, manned guided missiles.

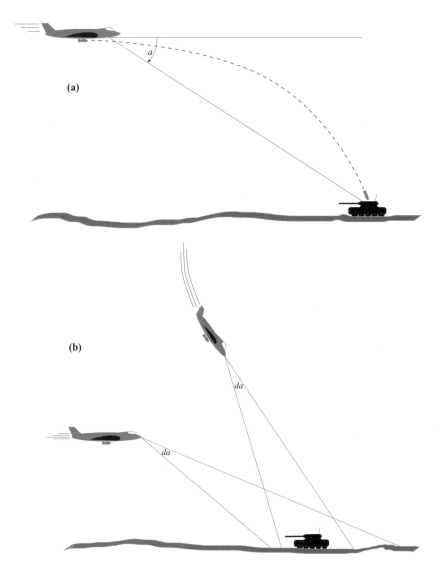

Figure 6.4. Area bombing and dive bombing. (a) Area bombing of a target requires predicting the flight of a bomb released from an airplane; because the line of sight differs considerably from the bomb trajectory, there is room for considerable error. (b) Dive bombing leaves less room for error. If the estimate of target direction has an angular error of *da* degrees, this translates to a smaller area on the ground than the same angular error for an area bomber.

Figure 6.5. A Douglas Dauntless dive bomber over Wake Island in October 1943. U.S. Navy photo.

highly successful in the battles of the Coral Sea, Midway, and Guadalcanal. This two-seater aircraft was well armed and could defend itself against Japanese fighters. In the last two years of the war, the Dauntless was replaced by the much faster Curtiss Helldiver.

ROCKETS COME OF AGE

In chapter 2, I mentioned in passing the military use of rockets by twelfth-century Chinese warships. Not much progress was made in rocketry from then until the eighteenth century, when an Indian army of the Mysore Wars dispersed their British enemy in the Battle of Guntur (1780) by firing rockets at them. William Congreve, a British Army officer, learned from this experience (a number of Indian rockets were sent back to England for analysis), and he produced rockets that were later used by the British in the Napoleonic Wars. Congreve rockets were also used against Americans in the War of 1812 (as reflected in Francis Scott Key's line—"The rockets' red glare . . ."—written after the bombardment of Fort McHenry). These missiles consisted of a cylindrical iron case with a guidance stick (up to 15 ft long) emerging from the center of the case, powered by black powder. Congreve rockets were capable of reaching an altitude of 9,000 feet, but their use as a military weapon was debatable because of their poor accuracy (due, no doubt, to poor and inconsistent manufacture). Thus, while 25,000 rockets were usefully deployed against the city of Copenhagen in 1807, any smaller target was reasonably safe (or, at least, in no more danger than the soldiers who ignited the rockets).[8]

Another English rocket inventor was William Hale. His rocket, invented in 1844, had no guide stick and so, unlike the Congreve, was aerodynamically unstable. Subsequently, the Hale rocket, which was adopted by the U.S. Army in the middle of the nineteenth century, was provided with fins that caused it to spin, producing gyroscopic stability. During this period rockets began to be of practical use; rocket-powered harpoons were developed for whalers, and rockets carried rescue lines to ships foundering near shorelines. The first rocketry theory appeared at the turn of the twentieth century. Konstantin Tsiolkovsky, in Russia, showed that the performance of a rocket was limited by the speed of the exhaust gas.

Perhaps because of the dominance of artillery in World War I, rockets seem to have been used very little during that conflict. Rocketry took off, as it were, in the interwar years in America and Germany. Robert Goddard developed the first liquid-fuel rocket in 1926, and in 1934 Werner von Braun wrote his doctoral thesis on the subject (the thesis was kept secret by the Nazi authorities).

It was during World War II that large, medium-range rockets first hit the streets—literally. The Peenemünde Research Facility, on the Baltic coast of Germany, produced the V-1 flying bomb and the V-2 rocket, both "vengeance weapons" deployed during the last year of the war by the increasingly desperate Nazi regime. The V-1 (known in England as the "doodlebug") was powered by a pulse-jet engine that gave this missile its characteristic buzzing sound. It had stubby wings and flew to its target with the aid of a simple autopilot that controlled height and speed. The V-1 was, in effect, an early cruise missile. It moved slowly enough for a fast fighter to position itself alongside and then tip over the missile with its

8. The Congreve rocket also played an important part in the Battle of Leipzig in 1813, and a year later against the Americans in the Battle of Bladensburg, outside Washington, DC.

wing. Several thousand V-1s were fired at London and later at Antwerp, though only about a quarter of them got through to their targets.

The V-2 was altogether more effective. This was a true ballistic missile, descending on its target city at a high speed (M4) from the edge of space. This approach made the V-2 invulnerable to fighter defense and anti-aircraft fire. Designed by von Braun, the liquid-fueled V-2 had a range of 225 miles and was armed with a 1,650-pound warhead. Between them, the V-1 and V-2 caused the deaths of about 9,000 people in their target cities (plus about 20,000 deaths among the labor camp inmates used to manufacture them). A more potent V-3 rocket was in the experimental stage by the time the Peenemünde facility was overrun by the Allies.

Many of the German rocket scientists who had worked at Peenemünde were transported to the United States after the war to a research facility in Huntsville, Alabama, where they continued where they had left off "under new management." One early result, in the 1950s, was the Red-stone rocket, a liquid-propelled surface-to-surface missile. Thereafter, large-scale rocketry became increasingly devoted to the space race and to guided missiles, which are not part of our story, and to intercontinental ballistic missiles, which are (see fig. 6.6).

Small-scale rocketry also made a big impact in World War II. Again, the lead developers were the Americans, with their bazooka,[9] and the Germans, with their Panzerfaust. Both of these weapons were early *rocket-propelled grenades* (RPGs). They fired a shaped charge, of which more in chapter 7, which was devastating to tanks. The Panzerfaust formed the basis for the later Russian RPG-7, variants of which have found widespread use to the present day (fig. 6.7).

Another example of a World War II surface-to-surface rocket was the Soviet Katyusha ("Little Katie"). These short-range (about 3 miles) solid-propellant missiles were inaccurate,[10] but this did not matter, as they were a barrage weapon, intended for blanket coverage of enemy positions. The

9. The name came from its resemblance to the musical instrument. The idea behind bazookas came from Goddard during World War I. See Hacker (2006, pp. 98–100).

10. Solid-propellant rockets were less accurate than liquid-propellant rockets because of irregularities in the fuel, which led to an inconsistent burn and erratic flight. For a good introductory account to the evolution of rockets, see Van Riper (2007).

inexpensive Katyushas were used by the thousands and were much feared by the Germans. Katyushas were launched from trucks; a truck could hold up to 48 launchers. Because launching was quick, the trucks could redeploy rapidly before attracting enemy fire.

Air-to-ground rockets also came of age during World War II. One example was the British RP-3 rocket, launched from under the wings of the Typhoon ground attack aircraft (fig. 6.8). As with all rockets of the day, accuracy was a problem. The 60-pound high-explosive warhead was very effective against soft targets such as trucks, trains and rail yards, boats, and U-boats but less effective against tanks, unless they struck a track or engine.

Rocket Science

You may regard a rocket as an artillery shell with a very-slow-burning propellant and ask, "What is the big deal?" Why should rockets change the face of warfare, both tactically and strategically? The differences between

Figure 6.6. The launch of a LGM-30 Minuteman III intercontinental ballistic missile. These missiles are guided only during the launch phase; most of their trajectory is unguided—ballistic. The problems involved in aiming ICBMs are similar to those for the Paris Gun, discussed in chapter 5, except more so: the range of ICBMs is several thousand miles. U.S. Air Force photo.

Figure 6.7. An Iraqi Security Force commando shoulders an RPG-7. Note the biconical-shaped charge munitions. Photo by LCpl. Kenneth Lane, U.S. Marine Corps.

rockets and gunpowder munitions are easy to see, and the ramifications of these differences are profound. Let me explain.

Rockets of the nineteenth century were at best a useful adjunct to a military campaign. They were an unreliable novelty that was successful at frightening horses and even hardened soldiers, but they rarely hit their intended targets. Later, improved manufacturing and better understanding of how rockets work led to improved accuracy, though still not nearly so good as the accuracy of a bullet or artillery shell. The main benefit of rockets, which became evident during World War II and thereafter, is that they impart little or no recoil. This does not mean that Newton's third law does not apply to rockets, but instead means that the reactive force is directed differently. An accelerating bullet "pushes off" the chamber of a rifle barrel, which then recoils against the shooter's shoulder. An accelerating bazooka round pushes against the gases it expels out of the back of the bazooka. The soldier who fires the bazooka feels next to no force: this is the great advantage of rocket weapons.

Let's think about this. Those Katyusha rockets were launched from light trucks. Each rocket had a 44-pound warhead. An artillery weapon firing

Figure 6.8. Sixty-pound air-to-surface rockets being loaded aboard a World War II Typhoon aircraft. Image from Wikipedia.

shells with the same explosive charge, to the same range, would have recoiled so much it would have destroyed the truck. Those Typhoons each carried rockets with 60-pound warheads under their wings: any airplane carrying an artillery piece that fired 60-pound shells would break up because of the recoil. Rockets place much less of a demand on the launcher: the launch frame can be much lighter (and therefore less expensive and more mobile).

Here is a telling example. Most World War II Soviet tanks could be destroyed by a German 88—the famous 88-mm anti-aircraft gun, which was also widely used against armored ground targets; many tanks could also be destroyed by a Panzerfaust rocket. A 5-ton artillery piece versus a man-portable rocket launcher. (Not entirely a fair comparison, of course. The problem with rockets from this period, as we have seen, was their accuracy. This lack of accuracy restricted rocket use on the battlefield to short-range or barrage weapons.) The missile projectile itself can be less robust if it is powered by a rocket motor instead of a propellant charge. A bullet emerging from the muzzle of a modern rifle is under a 100,000-g

force;[11] therefore, it—and the rifle—have to be physically strong. The same speed can be obtained with much gentler forces via a rocket motor.

Lack of recoil means lighter platforms as well as more punch for a given weight of material. This trend has continued beyond World War II to the present day. Nowadays, should you want to bombard Paris from 75 miles away (I would discourage you from any such thoughts—Paris is truly one of the great cities of the world), you would no longer need a 125-ton gun especially designed for the purpose and an 80-man crew. V-2s would do the trick with comparable accuracy and a bigger warhead from farther away; modern ballistic missiles would be more accurate from much farther away. The improvement in ballistic delivery is due to rocketry. (The improvement in accuracy required something more—the development of remote sensing and sophisticated guidance systems.)

The physics of rocket ballistics is interesting because the mass that is being moved through the air (or through space) changes with time, as propellant gas is expelled. This feature makes rocketry basics the stuff of university physics courses. You will find in technical note 20 the rocket science that lies behind figure 6.9. This graph needs some explanation—but it is worth the effort to understand because it tells us quite a lot about how rockets work and how they compare with artillery rounds that have the same payload mass and the same mass of propellant. If you are following the math, all the details are in technical note 20; if not, read on.

One of the standard exercises in undergraduate mechanics is to show how the mass of a rocket falls as the rocket picks up speed. This is a consequence of momentum conservation if our rocket is out in space where there is no aerodynamic drag. For such a case, let's assume that the rocket starts at zero speed; as it expels gas, it accelerates. Thus, speed increases and mass decreases as plotted in figure 6.9a. For us, however, this exercise is not good enough because our rockets are not in space; most often they are on a battlefield near the earth's surface, where they are subjected to aerodynamic drag.

Accounting for the effects of drag makes the calculation more complicated, but it can be done (see fig. 6.9a). We saw earlier that drag imposes a maximum speed on a projectile: it cannot keep accelerating to ever-higher speeds. This is true for airborne rockets as well as for bullets, except

11. This figure is readily calculated from a knowledge of muzzle speed and barrel length, plus a few reasonable assumptions.

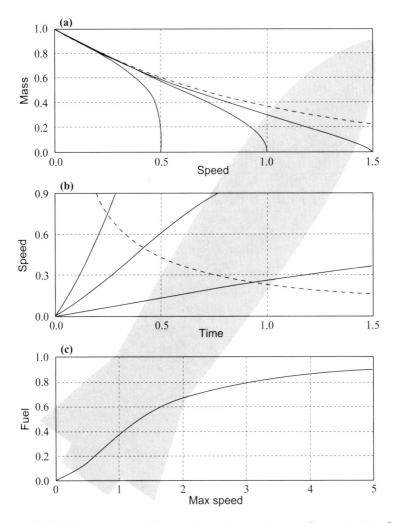

Figure 6.9. Rockets vs. gunpowder munitions. Here trajectory characteristics of a rocket and a matching artillery shell are compared. (a) Mass of a rocket vs. speed, in space (dashed line) and in the atmosphere (solid lines) for three maximum speeds (0.5, 1.0, and 1.5 times the gas ejection speed). (b) Speed vs. time for a shell (dashed line) and for rockets at three maximum speeds, as in (a). For this calculation, the fraction of initial rocket mass assumed to be taken up by fuel is three-fourths. The rocket can always be made to be faster than the shell. (c) The fuel fraction required to ensure that a rocket is as fast as a matching shell, vs. speed. Above the line, rockets are faster.

that rocket engineers or the rocket pilot can select the maximum speed that they want, within limits. This is because a rocket's maximum speed through air depends upon the rate at which gas is being ejected as well as upon aerodynamic drag coefficient. Three of the curves plotted in figure 6.9a correspond to three different maximum speeds. As the rocket accelerates toward the chosen maximum speed, the curves approach the curve for a rocket in space. Say that the mass of a rocket plus its payload (the warhead, for example) is 20% of its initial mass (with the other 80% being fuel, so that the fuel fraction is $\epsilon = 0.8$). Figure 6.9a shows how close to its maximum speed this rocket can get before running out of fuel. Here we have exposed one of the factors that rocket engineers need to account for when thinking about ballistics: how much fuel to supply the rocket (be it a bazooka round or a Minuteman ICBM) so that it will reach its desired speed given the required payload.

Figure 6.9b shows how the speed of a rocket changes with time, from the instant that it is fired, assuming the same three maximum speeds as before. Also shown is the speed of a matching shell—one that has the same mass as the rocket payload and the same mass of propellant (so that rocket fuel mass is set equal to shell propellant charge mass). Because of drag, the shell speed falls inexorably with time once it has left the gun barrel. The rocket can accelerate after leaving its launcher, since it still has plenty of propellant on board. The key feature to note from figure 6.9b is that the rocket speed can always be made to exceed the shell speed, whatever the chosen maximum shell speed may be. It may be necessary to adjust the fuel fraction, ϵ, in order to achieve this result. In figure 6.9b I have assumed that $\epsilon = 0.75$, so that three-quarters of the initial rocket mass is fuel.

To emphasize the speed advantage of rockets, I have plotted in figure 6.9c the fuel fraction ϵ required so that the maximum speed of a rocket equals the speed of a matching shell.[12] If the rocket is given a higher value of ϵ then it will impact the target at higher speed. You can see that, especially at low speeds, it is easy to make rockets faster than matching shells.

12. Here I am referring to the speed of the shell when it hits its target and the speed of a rocket hitting the same target. I assume that the rocket runs out of fuel at this instant. For ballistic missiles this is a reasonable assumption. Many guided missiles, however, operate on a boost-coast strategy: they run out of fuel before reaching their target and coast the rest of the way. This extends range and enhances maneuverability because with no fuel, the rocket is lighter and easier to maneuver. You'll find more on guided missiles in the last section of this chapter.

ICBMs and Space Rockets

Ballistic missiles are fired from the earth's surface (or below the ocean surface in the case of submarine-launched ballistic missiles, or SLBMs). Those of medium range follow a near-parabolic trajectory that takes them up into the stratosphere before returning to earth. This means that they are subjected to the same influences as were the shells of the Paris Gun: the forces of gravity, aerodynamic drag, and twist; Magnus and Coriolis forces; and the significant effects of atmospheric attenuation with altitude. Ballistic missiles are rocket powered only at the beginning of their trajectory and are unguided, so we can regard their ballistics, in this book, as pretty similar to those of a long-range shell. For even longer-range missiles such as ICBMs these forces and effects are more apparent, but the basics remain the same.

The only additional consideration here is the multistage rocket such as those used to deliver a payload into space; the use of multistage rockets also makes sense for ballistic missiles with an intercontinental range. Long-range rockets require a very large fuel fraction, ϵ, and this fuel must be contained within the rocket body. (Think of a space module sitting atop a giant Saturn launch rocket at Cape Canaveral.) It makes sense to ditch the fuel tank once the fuel has been consumed—no point in lugging around all that extra mass. Calculations show that this simple logic is sound: the speed of a two-stage rocket, which ditches its first-stage fuel tank when empty, is greater than that of a one-stage rocket with the same amount of fuel. The only detail I need add is that the benefit of a multistage space-bound rocket must occur at low altitudes and speeds so that the jettisoned fuel tanks fall back to earth. If the tanks are jettisoned at high speed—high enough to take them into orbit—energy has been wasted; you may as well have left them attached.

SEEKING GUIDANCE

Guided missiles are powered throughout their flight: they have to be, so that they can respond to navigation input and make course corrections. So, rockets are suitable as propulsion for guided missiles as well as for ballistic missiles. Jet engines also power guided missiles.

You might regard the old V-1s and V-2s as guided, at least crudely, because their ballistic trajectories were interrupted by engine cut-out, instigated by a timer when the missile was over its target. A simple instruction ("cut engine after flying for *x* minutes") resulted in a course change. Now multiply up this instruction set a million-fold, and you have the basis of a modern guidance system. Such systems are beyond the compass of this book because we are here interested in ballistics and thus only in the trajectories of unguided projectiles. Nevertheless, for completeness I feel that I owe you some explanation of missile guidance principles, if only to place in historical and engineering context the older and more common projectile trajectories.

A more sophisticated version of the V-1 missile guidance instruction might read like orienteering instructions: "Fly due N at 400 mph for 17 minutes and then turn NE and fly at 600 mph; then dive vertically at maximum speed." This type of guidance instruction is known as *open loop*, meaning that there is no feedback to correct for any errors that arise during the flight. For example, headwind may slow down the ground speed of the missile, so when it dives onto its target it is not hitting the location intended. No modern guided missiles use open-loop navigation and guidance systems. The *closed-loop* systems in place today incorporate feedback from the environment to correct for any errors in trajectory position and velocity that arise. Such systems require remote sensing capability: the missile has to be able to receive information from outside.

In the smallest and therefore simplest guided missiles, the information received might be simply a GPS coordinate, updated frequently so that the missile can turn toward the target coordinate at all times. A *smart bomb* receives another type of target location indicator; perhaps the target is illuminated by a laser beam,[13] so that the bomb maneuvers by adjusting its control surfaces (fins or winglets) to ensure it is heading toward the right location. *Beamrider* missiles follow a beam (the pencil beam of a millimetric radar transmitter or a laser beam) that is pointed at the target for the duration of the missile trajectory. By riding down the beam, the missile will get where it is meant to go. Beamrider systems were commonly

---

13. To a radar systems engineer, "illuminate" means to reflect energy from the target (in this case laser beam energy); it does not mean that the target is lit up by laser light. (In a radar context, it may mean to bathe the entire target in microwaves, which then reflect back to the radar receiver.) A more colloquial term meaning the same thing is to "paint" the target. The choice of microwave (radar) or IR/optical (laser) for the illuminating beam frequency is not always obvious. Laser beams are narrower, permitting greater accuracy, but unlike microwaves they are susceptible to scattering and attenuation by fog, clouds, or rain.

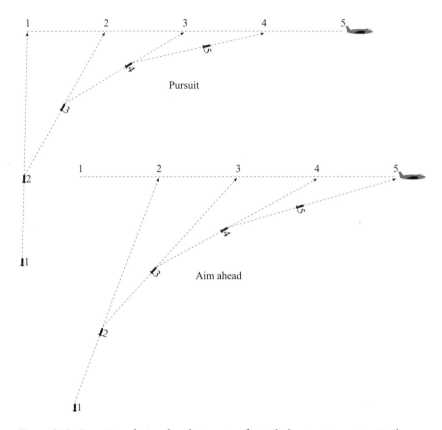

Figure 6.10. Pursuit and aim-ahead strategies for tail-chasing air-to-air missiles. The simplest strategy is to point the missile in the direction of its target. This is the pursuit curve of a dog chasing a rabbit. Better to aim ahead and direct the missile to where the target is going to be. Aim-ahead capability requires of the missile a more sophisticated sensor system and clever computer algorithms.

provided for air-to-air missiles. On the ground, short-range antitank missiles may be *wire-guided*. Here, a thin wire connecting missile to launcher is spooled out as the missile flies; electronic instructions from the missile crew are sent down the wire to direct the missile on target. In all these examples, the missile crew is actively guiding the projectile onto its target.

Longer-range and more modern missile guidance systems are of the *fire-and-forget* type. The oldest and simplest of these are the heat-seeking air-to-air missiles that follow the plume emanating from the jet of an enemy airplane. Such tail-chasers are provided with an infrared (IR) sensor in the nose cone that provides target direction information to the missile, which then adjusts course as necessary to catch up with the (perhaps frantically maneuvering) target aircraft. The tactical advantages of fire-and-forget missiles in a dogfight are obvious: the pilot need not spend time directing his missile; instead, he can reposition his airplane for another attack or maneuver to avoid an attacking enemy. Heat-seeking missiles are unsophisticated by modern standards, but even so, missile engineers invested a considerable effort into optimizing pursuit strategies. A simple example is shown in figure 6.10.

An early type of cruise missile was the sea-skimmer, developed during the Cold War decades. These ship-to-ship or air-to-ship missiles could be fired from over the horizon; they approached their target at high speed just a few feet above the waves (the altitude could be adjusted to suit sea conditions), beneath target radar, and homed in on the target using information either from their own nose cone radar sensor or from the target ship's radar. True modern cruise missiles have a much more sophisticated navigation and guidance system, enabling them to make intelligent decisions in changing circumstances. For example, a cruise missile might be provided with a preprogrammed map database of the battle space area as well as GPS target coordinates; it can make course corrections by comparing what is seen by its imaging radar sensor with what its database says should be there. A cruise missile can make decisions to take a less direct route to the target to avoid bad weather, say, or enemy action. Cruise missiles are programmed to pick out, for example, a particular building in a particular enemy city hundreds of miles away, or a particular entrance of an underground enemy installation.

Unmanned air vehicles (UAVs, more politically correctly referred to as "uninhabited air vehicles") are remotely operated aircraft without crew. These vehicles are thus smaller, lighter, and less expensive than a conventional airplane; they do not need to accommodate the weight or life-support systems of a crewed airplane. UAVs are equipped with several high-resolution sensor systems—optical, IR, radar—and can carry munitions. They are remotely operated by a pilot at a safe distance—perhaps hundreds of miles away. Not missiles, but missile platforms, these are probably the way of the future. Next-generation UAVs (some prototypes are with us today) will operate autonomously: not even a remote pilot will be needed. They can just be told to go into enemy territory and do something.

In a hundred years' time, I foresee *smart bullets* that will enable a hunter to fire his rifle out of his den window, seek out a deer, kill it, and transmit coordinates back to the hunter. Hardly seems fair, really.

Improved manufacturing techniques led to firearm range increases and enabled the development of fire control equipment to guide artillery shells and bombs. In World War II dive bombers became very important because they were the most accurate method of bombing. Rockets also evolved into effective weapons during this period. The main tactical advantage of rockets is that they are recoilless; they could thus be launched from a lighter platform than was required for a conventional ordnance round with the same payload. Also, a rocket can be made to fly faster than an equivalent artillery round. Guided missiles solved the ballistic missiles' aiming problem (increasingly difficult as ranges got longer). The gentler acceleration of rockets permit much more sophisticated onboard navigation and guidance systems than is possible for missiles launched from guns.

Part III

THUD!
Terminal Ballistics

7 Stopping the Target

The purpose of a projectile weapon is, and has been since prehistory, to stop a target. To stop it from moving, from being able to hurt you, or from breathing. This purpose is not necessarily achieved simply by hitting the target; a sling stone may bounce off a leather jerkin, or a cannonball may bounce off the wooden sides of an Age of Sail ship.[1] If this happens, then the projectile has failed in its purpose: no matter how skillful the shooter has been in launching and directing his missile to the chosen target, if the missile does not damage or incapacitate the target, then the whole exercise was a waste of time. As a consequence, since prehistory people have given thought to the effects of a projectile upon the target it hits and how these effects can be maximized. A sling stone made of lead may have a greater impact on an enemy soldier than one of stone, even through a leather jerkin. Heavier or faster cannonballs may have a more destructive effect on the hulls and masts of enemy ships.[2]

There may be more than one acceptable outcome from a ballistic projectile strike. Age of Sail ships were valuable prizes, so you, the captain of a frigate out to make a name and fortune for yourself by capturing enemy "prizes," will destroy the enemy only if you cannot capture him. You do not want to smash his hull and sink him; you make use of specialized

1. The USS *Constitution*, a "super frigate" constructed by the young U.S. Navy especially to combat the smaller British frigates during the War of 1812, was built from live (evergreen) oak and was resistant to the weight of the British cannonballs. The balls simply bounced off, and this venerable ship (the oldest commissioned warship still afloat) was consequently dubbed "Old Ironsides."

2. It is reported that an early-nineteenth-century 32-pound cannonball (the gun itself weighed over a ton and a half) could smash through 3 feet of solid oak at close range (100 yards). See, for example, Heath (2005, chap. 12). Many of the injuries to ships' crews came not from the ball itself but from the high-speed splinters that were sprayed over the decks.

ammunition designed to achieve other purposes. Grapeshot (early shrapnel; think of a supersized shotgun cartridge) was used to sweep the decks of enemy crew while leaving the ship relatively undamaged; chain shot was used to bring down rigging and damage sails, thus hindering maneuvering by the target ship. Earlier in history, arrows were provided with different heads for different purposes. Flat arrowheads for penetrating flesh were given barbs so that they could not easily be pulled out. The deep wounds caused by an arrow could be made worse by dipping the arrowhead in dung before firing it, leading to infection. (A hunter after game would not use this tactic; it was suitable only for human prey.) Armor-piercing rounds, to use a modern phrase, took the form of conical *bodkins* that could punch through mail or plate armor.[3]

The modern period of gunpowder weapons has seen a proliferation of warheads and of scientific investigation into their penetrating and incapacitating power. The study of this last phase of a projectile's trajectory, dubbed *terminal ballistics*, is a sufficiently mature discipline to have split into several specialized fields of study. In this chapter we will look at two of the main areas of interest: the terminal ballistics of a projectile (a specialized armor-piercing round) penetrating a metal target, such as a tank hull or turret, and the terminal ballistics of a bullet entering a human or an animal. The first area is part of the arms race that has been going on since World War II between armor-piercing weapons and armored vehicles; the second area is directed toward understanding the physiological effects of a bullet wound and their causes. In both cases the results of extensive study contain a few surprises and are of great interest to anyone who is curious about ballistics.

HITTING METAL

Throughout the years of World War II, tanks became heavier and heavier in response to antitank guns' becoming larger and larger. It was found, as you might expect, that harder, tougher, heavier and more oblique monolithic plate armor is more resistant to high-explosive (HE) or solid-shot artillery rounds than softer, less tough, thinner steel plates that are presented face-on to the projectile trajectory. This is why tanks became heav-

---

3. There are many books about Age of Sail warfare; see Nicolson (2005) for a particularly vivid account of the Battle of Trafalgar. See, for example, Hall (1997, chap. 4) or Denny (2007, chap. 1) for a description of the military use of archery.

ier and with more angled plates. You may think that the guns would quickly win this race because there are obvious limits to the weight of a moving vehicle. Not so, however, because large antitank guns were very immobile, and this factor made them vulnerable to tanks and to enemy artillery in general. Towards the end of the war, a different kind of armor-piercing round came to the fore, the rocket-propelled grenade. We met RPGs in chapter 6; now it is time to look at the way that their shaped charges were able to punch through monolithic steel plates to disable tanks and other armored vehicles.

Shaped Charges

Figure 7.1a shows a World War II German soldier armed with the Panzer-faust antitank weapon, so effective against Soviet tanks during the final months of the war.[4] The biconical shape of these warheads is seen in their modern equivalent, the shoulder-launched RPG (shown in fig. 6.7). There is a reason for this form, and it is connected with the shaped charge that provides the lethal punch.

In 1888 the American Charles E. Munroe, working for the U.S. Navy, discovered that a blast of high explosive could be made to concentrate or focus in a chosen direction by a suitable choice for the geometrical shape of the charge. This phenomenon is now known as the *Munroe effect* or the *hollow-charge effect*. Thus, a block of HE with a V-shaped groove cut in it would, when detonated, emit a blast that was concentrated in a plane emanating from the groove, instead of expanding in all directions. This is the basis of the linear-shaped charges used today by building demolition engineers to slice through structural steel columns. Imagine, now, what happens when the V-groove is replaced by a conical hole in the block of HE: the result is a blast that is concentrated along a line emanating from the hole. Because the blast is concentrated in this way, any material that is caught up in the blast is propelled to very high speeds and forms a jet.

The basic design of a shaped charge round is shown in figure 7.1b. The conical hole is lined with metal—usually copper—and when the charge

---

4. To be fair, the Panzerfaust was effective against most tanks, not just Soviet ones, but the majority of tanks that the Axis powers had to combat during these final months of the war were Soviet. Also, the proliferation of Panzerfaust RPGs had more to do with their low cost than their effectiveness. The best antitank weapon of World War II was another tank, but the collapse of German armaments production capability at the end of the war limited Germany's options.

(a)

(b)

Figure 7.1. World War II rocket-propelled grenades. (a) An early Panzerfaust RPG, here in the hands of a Luftwaffe soldier. Note the conical warhead. (b) Cutaway illustration of a shaped charge projectile.

detonates, this liner disintegrates, joining the jet. The pressure generated along the jet can reach 45 million psi, which accelerates the small particles of metal in a thin stream directly forward. Only about 20% of the metal liner forms this stream; the remaining 80% forms a relatively slow moving plug which follows on behind. The stream head reaches an astonishing speed of 8–9 km s^{-1} while at the tail the speed is about 1 km s^{-1}. The plug trundles along behind at 300 ms^{-1}.

How does the stream of very-high-speed copper particles punch through steel armor? Very easily. At the front of the stream, the particles have so much energy and momentum that steel armor looks like a fluid rather than a solid, and the stream just pushes it aside. One important feature of the shaped charges is that, to be effective, the charge must detonate a little distance in front of the armor. This (and aerodynamic streamlining) is the main purpose of the conical head. The charge detonates when the head comes into contact with the armor, and the empty space between charge and nose cone tip is enough for the copper stream to form.

High-Explosive Antitank Rounds

High-explosive antitank (HEAT) round is the modern name given to the shaped-charge ammunition fired by Panzerfausts and bazookas. Since World War II the effectiveness of these rounds has improved a lot. They are devastating to the crew of a stricken tank because once the copper stream has punched a hole in the tank armor, metal particles (including *spalling* from the inside surface of the armor) spray the crew and can ignite ammunition. Modern HEAT rounds are fired from smooth-bore barrels because it was found that gyroscopically stabilized HEAT rounds are less than half as effective: centrifugal force acting upon the copper stream causes it to disperse. As a consequence, modern rounds are fin-stabilized.

Today, HEAT rounds can be fired from artillery as well as shoulder-launched infantry weapons; they can also be rocket-powered guided weapons. These developments mean that HEAT rounds now have longer ranges and higher trajectory speeds. The broad conical nose cone has been replaced by a much thinner tube, which acts as the spacer to ensure detonation at the correct distance from the target armor. At long range, guided missiles are the preferred weapon against *main battle tanks* (MBTs) and other armored vehicles. The light weight of rocket launchers and the small size of HEAT munitions mean that helicopters have become the deadliest antitank weapons platform. A helicopter can carry many HEAT rounds and can direct them from 4 km away down onto the top of an enemy tank, where its armor is weakest.

Armor Reacts to HEAT

The Munroe effect and low cost have made HEAT rounds very effective weapons against tanks. The earliest responses to these weapons by tank crews was ad hoc and dates from World War II. There is a grainy old photograph of a Soviet tank during the Battle of Berlin in 1945 equipped with a hastily added metal cage placed along the side of the hull. (It is reported that Soviet tank crews resorted to covering their tanks with metal-spring mattresses looted from Berlin households to counter the Panzerfaust threat.)[5] In the Vietnam War, American tank crews would enhance their defenses by adding sandbags to the hull and turret. Nowa-

---

5. See Beevor (2002) for an authoritative and very readable account of the Battle of Berlin and for the report of mattresses used to defend tanks.

Figure 7.2. Slat armor on a Stryker combat vehicle. U.S. Army photo by Jim Hinnant, 401st Army Field Support Brigade.

days tanks and lighter armored vehicles may have *spaced*, or *slat*, armor, a more permanent and robust version of the earlier metal cages but using the same basic idea (fig. 7.2). Although cages, sandbags, and slat armor are not substantial enough to protect against a conventional HE round, they are effective against HEAT rounds because they cause premature detonation. The copper stream forms too far away from the steel armor and becomes dispersed, so that it loses much of its penetrating power.

Or that is the theory, but in practice this hiding-under-a-mattress approach does not always work. Tanks are very expensive, and HEAT rounds are not, so combatants faced with enemy tanks can fire large numbers of HEAT rounds and win, even if the success rate of an individual round is low. Tanks were losing the battle against HEAT missiles by the 1960s and 1970s, but tank designers responded effectively in two ways. The first approach, Chobham armor, added significantly to the burgeoning cost of MBTs, while the second, reactive armor, is an eloquent statement of the desperation to which tank designers were driven.

Chobham armor is a multilayered metal-ceramic armor that was first developed in the 1960s in England (Chobham is the English village nearest the research facility where this armor was invented) and later in the United States. It consists of very hard ceramic tiles encased in a steel matrix. The combination of hard, brittle ceramic with tough, ductile metal is difficult for the high-speed particle stream from HEAT missiles to penetrate—there are lots of surfaces to deflect the stream. Additionally, modern Chobham armor is backed by steel plate to hold the composite structure together and has shock-absorbing elastic layers. The M1 Abrams (fig. 7.3), the first MBT to be encased in Chobham armor, it is said to have been proven effective during the first Gulf War.

Chobham armor works best when an incoming antitank missile strikes it perpendicularly rather than at an angle. This is counterintuitive and goes against our experience with monolithic metal armor. We have already seen that the tanks of World War II were given sloped armor; this provided an increased thickness of steel to protect against nearly horizontal artillery shell trajectories and also perhaps served to deflect a solid-shot round. Chobham armor gives a tank a quite different look: note the steep sides of the Abrams. Many of the rules for tank armor design noted earlier (heavy, thick, oblique armor is better . . .) do not apply to composite armored vehicles. Chobham armor weakens after it has been struck. Thus, even though the tank and crew survive a hit that would have destroyed one of

Figure 7.3. An M1 Abrams main battle tank. This tank weighs 61 tons, and its main armament is either a 105 mm rifled cannon or a 120 mm smoothbore cannon. U.S. Army photo.

the previous generation of tanks, they are vulnerable to a second hit in the same place.

Nowadays the hulls of tanks and other armored vehicles are given a Kevlar lining to protect crews from the spalling of the inside surface of a metal hull struck by an antitank round. Lighter vehicles are made from metals other than steel; for example, titanium is used for lightweight armored vehicles, while an aluminum-magnesium alloy is now common for U.S. armored personnel carriers and howitzers.[6]

Reactive armor consists of a jacket made up of explosive bricks that covers the metal hull and turret of a tank. Each brick is a sandwich of explosives between metal plates. Using explosives to defend against an HE or HEAT round sounds crazy, doesn't it? The idea is that an incoming HEAT round would cause a reactive armor brick to explode; this explosion would be too small to damage the tank but would be enough to deflect the high-speed jet that HEAT rounds rely upon for penetrating armor. Reactive armor was invented independently in West Germany and the Soviet Union in the 1960s. Since then, a number of variants have been developed.

The type described here is called explosive reactive armor, predictably enough. It works quite well, but has two fairly obvious problems. First, it is often a good idea, tactically, to have troops accompany tanks into battle—but exploding reactive armor may result in casualties among one's own infantry. Second, reactive armor is even more vulnerable than Chobham armor to a second hit in the same place. Designers of antitank missiles know this, and so they have come up with a *tandem charge* HEAT round. The first charge sets off the reactive armor on the tank; the second charge then sends the copper jet through the second layer of metal armor.

Antitank Arrows

In recent decades another arrow has been added to the quiver of effective antitank rounds, taking the form of the *long rod penetrator* (LRP), a kinetic energy weapon that relies on mass and speed, not chemical energy, to burst through tank armor. Such weapons are cousin to the cannonballs of two centuries ago; when wooden sailing ships gave way to steam-powered ironclad warships, solid shot projectiles were retained, as the best bet for

6. For an introduction to tank armor, see Montgomery and Chin (2004). For more on the M1 Abrams, see Green and Stewart (2005). The introduction to Norris and Marchington (2003) provides a useful survey of antitank weapons.

punching a hole through the armor. Rather than being round, however, LRPs are arrow-shaped because this is the most effective form, for reasons that will soon be seen. Some simple physics (Isaac Newton's law of momentum conservation) is enough to paint us a broad-brush picture of what is going on here: it tells us that, in addition to being long and thin, the most penetrating projectile will be very dense and very fast (*hypervelocity* is the Star Wars term commonly found in the technical literature). The Newtonian argument is summarized in technical note 21. While approximate, this argument must be close to the truth because the most effective way currently known for penetrating thick steel armor is with LRPs; they are increasingly taking over from HEAT missiles at the top end of the market.

LRPs are constructed from either tungsten or depleted uranium, both of which are about 19 times denser than water (and twice as dense as steel). They travel through the air like greased lightning: LRP muzzle speeds vary between 1,400 ms^{-1} and 1,800 ms^{-1}—about one mile per second, or four or five times the speed of sound. A typical LRP is 2–3 cm in diameter (say, 1 in) and 50–60 cm long (20–24 in) and weighs about 4 kg (9 lb). I would not want to get in the way of one of these projectiles.

In addition to being more effective than HEAT missiles against tank armor, LRPs are more accurate, owing to their high speed. LRPs are fired from artillery guns, so how are such high muzzle velocities obtained for a small-caliber projectile? In fact, the artillery bore is much wider than the LRP diameter. Figure 7.4a shows a cutaway illustration of an LRP round; the shell casing is of much larger caliber (120 mm is typical) than the projectile. The LRP is held in place with a lightweight *sabot* (shown as dark grey in fig.7.4a) that is discarded during flight to reduce drag, as shown in figure 7.4b. Note that the LRP has fins. This is because it wastes energy to provide the LRP with spin; the round is fired from a smooth-bore gun and stabilized with fins.

As you might expect, there has been a great deal of experimental and theoretical investigation into the subject of armor penetration by projectiles—how to increase it and how to reduce it. Part of this research has resulted in several empirical formulas that summarize how far a given missile can penetrate a steel plate. The subject is technically complicated, and no fully developed theories yet exist; hence the need for empirical formulas, based upon experimental tests. One of these, Lambert's formula, is discussed in technical note 22. From this formula we can see

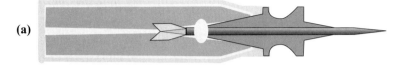

(a)

(b)

Figure 7.4. Long rod penetrators. (a) Cross section illustrating the LRP projectile, with fins and sabot attached inside the shell. (b) In flight, the sabot peels off the low-drag, hypervelocity projectile. Image from Wikipedia.

how fast two different types of NATO bullets need to be in order to penetrate steel armor to a given depth; the results are shown in figure 7.5. We can also play around with the formula and find out how far two different LRPs will penetrate steel armor (fig. 7.6). Note how effective they are: it is possible for one of these high-tech arrows to punch through half a meter of steel armor.

Finally, a word about the statistical nature of the empirical formulas. Such formulas arise from experimental tests, and the test results vary from sample to sample, leading to a distribution of missile speeds. Our formulas

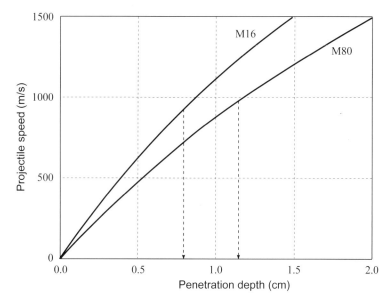

Figure 7.5. Speed required for two NATO rounds (fired from M16 and M80 rifles) to penetrate to a given depth of steel armor, estimated from Lambert's formula. For the actual rifle muzzle speeds (dashed lines) we see that the bullets will penetrate a little more or less than 1 cm.

estimate the speed, v_{50}, that must be attained in order for 50 out of 100 projectiles of a specified type to penetrate steel plate to a specified depth; thus, v_{50} is what statisticians call the *median* speed. If the same 100 projectiles are hurled toward the plate at a speed that is 15% lower, then perhaps only 10 of them will penetrate to the same depth; if their speed is 15% higher, then 90 will punch their way through.[7]

Tank leads to antitank weapon, which leads to tank armor designed to defeat the antitank weapon, which leads to improved antitank weapons—and so the arms race continues today. It seems ironic that the best modern antitank round harks back to the most effective projectile weapon of the fourteenth century—the arrow—if only in shape. The story is not yet over; this fight has several rounds to go (another pun—sorry). For what it is worth, my money is on the antitank weapons because they are inexpensive and because, to succeed, only one of them has to get through.

7. The statistical nature of penetration experiments is demonstrated in Zukas et al. (1982).

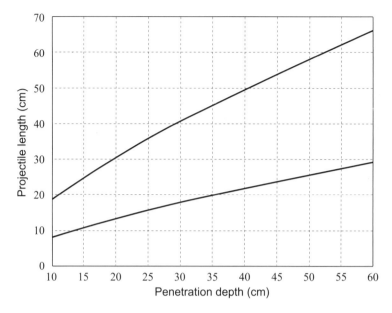

Figure 7.6. Projectile length vs. penetration depth for two LRPs fired at steel armor, estimated from Lambert's formula. The upper curve is for a 2-cm-diameter projectile striking the armor at 1,200 ms^{-1}; the lower curve is for a 3-cm-diameter projectile moving at 1,700 ms^{-1}.

HITTING FLESH

I will begin by knocking down the "knockdown" myth. Many people be-lieve that there is enough momentum in a bullet, even a handgun bullet, to knock a man down. This belief is bolstered, no doubt, by many of the less-cerebral action movies that show a bad guy getting shot while stand-ing still—and then he flies backward over a table or through a window. This is Hollywood, not reality. Yes, a person who has been struck by a bullet may double up, or suddenly twist or turn, but this is a physiological response; it is a twitch that can alter body shape but not move the body's center of mass across a table. Your getting shot in the chest with a .22 cal bullet when wearing soft body armor is equivalent, in terms of momen-tum transfer, to being hit in the chest by a baseball moving at 40 mph. It hurts—it leaves a bruise—but you are nowhere close to being knocked down. Let's up the ante by saying that you take a hit from a .45 caliber round: that is equivalent to being struck in the chest by a 90-mph baseball. You may hop about in pain, but you are not incapacitated or knocked down. Plenty of major league baseball players get hit by 90-mph baseballs without getting knocked over.[8]

Another example: the force from a 9 mm handgun bullet is about the same as that from a 1-pound weight falling from a height of 6 feet. Such a force is not going to knock you down. But suppose a more powerful rifle bullet did have sufficient force to push you over. Then Newton's third law tells us that the reaction force would also knock over the shooter. The movies get it wrong.

So is momentum transfer an insignificant factor in assessing the damage caused by bullet wounds impacting on human flesh? It seems that momentum *is* a key factor. The impact depth of a bullet correlates better with bullet momentum than it does with bullet kinetic energy. Kinetic energy transfer was an important consideration in penetrating steel armor, but the situation is not so clear in the (in some ways more complicated) assessment of wound damage. A bullet that passes straight through a person may cause less damage than a slower bullet that remains inside the body because the fast bullet transfers less momentum (and less kinetic energy) to the body.[9] On the other hand it may cause more damage, depending upon its shape and speed: as we will see, exit wounds can be much bigger than entrance wounds.

The correlation of a bullet's stopping power with its kinetic energy is uncertain. There are several theories of wound physics (a complex enough subject to warrant its own international organization, the International Wound Ballistics Association). Some assert that kinetic energy is the crucial component of ballistic wound damage, while others assert that shock waves are the real killer. To provide you with an early historical example that displays the difficulty in reaching a clear-cut (that's almost another bad pun) conclusion, consider again the Minié ball.

You may recall from chapter 3 that the Minié ball was the first bullet-shaped bullet; it made rifled muskets much more effective weapons than the older smooth-bore muskets. Civil War soldiers knew that the Minié ball

8. See Lee and Kosko (2005) for this baseball comparison. Patrick (1989) gives an FBI view on the knockdown myth.

9. Tissue damage occurs when energy is transferred from bullet to tissue, tearing it apart. On the other hand, momentum transfer causes tissue to be forced out of the way of a moving bullet. If a bullet passes right through a person and loses half its momentum in doing so, then it will leave behind three-quarters of its kinetic energy.

tore an enormous wound on impact and that an abdominal or a head wound would almost always be fatal.[10] Later, during the Spanish-American War, U.S. military doctors noticed that the wounds caused by the then-standard Krag-Jorgenson rifle were less damaging than wounds of 30 years earlier caused by a rifled musket firing Minié balls. Modern analysis confirms this observation. From a range of 10 feet, .58 caliber Minié balls were fired from a rifled musket into a block of ballistic gelatin (modeling human flesh). From the same range, full-metal-jacketed .30 cal bullets were fired into gelatin from a Krag-Jorgenson. The Minié balls weighed 458 gr and traveled at 944 ft/s, whereas the rifle bullets weighed 219 gr and moved at 1,852 ft/s. The Minié created a large temporary cavity in the gelatin (121 mm diameter), whereas the rifle bullet caused a 39-mm temporary cavity. Conclusion: the rifled musket would cause more severe wounds in human targets.[11]

Hmm. We have to be a little careful here before drawing conclusions:

- Ballistic gelatin is not human flesh. It does model the fluid-dynamical way that flesh responds to bullet penetration, but it can say nothing else about wound damage.
- Temporary cavity size—the maximum extent of the hole blown into a body—used to be considered the main indicator of wound damage, but now researchers are pretty sure that it isn't. We will see that there are many other contributors to wound damage in human tissue.
- The momentum of the two types of bullets is about the same. The kinetic energy of the .30 caliber bullet has twice the kinetic energy of the Minié. How can these facts be reconciled with the damage observed?

The lesson to take away from this introduction to ballistic wounds is that it is not easy to interpret results and come to firm, cut-and-dried conclusions.

Types of Wound Damage

There are a few general observations that can be made concerning the nature of bullet wounds. First the observations, made appropriately as bullet points, and then the discussion:[12]

10. Such a bullet wound may not have caused instantaneous death, but the damage was enough to lead to infection which, in the days before antibiotics, killed more soldiers than did gunfire.

11. This interesting experiment is reported by Dougherty and Eidt (2009).

12. Much of this section comes from Carlucci and Jacobson (2008), Di Maio

- Bullets from rifles are more damaging than those from handguns.
- At the same velocity, large-caliber bullets are more damaging than small-caliber bullets.
- Expanding bullets are more damaging than non-expanding bullets.
- Inelastic organs are more susceptible to bullet damage than elastic organs.
- Pressure waves or shock waves that emanate from high-velocity bullets entering the body can cause significant damage at some distance from the wound site.

There are four components of projectile wounding, and the above observations fit into one or more of these components: temporary cavity, permanent cavity, penetration, and fragmentation. A temporary cavity results from the fluid dynamics of a projectile entering a body—either a human body or a block of gelatin. The large hole that results from kinetic energy transfer lasts a brief period of time before closing over. Due to the elasticity of organic tissue, the temporary cavitation effects are not long term and are usually not fatal. However, some human organs are less elastic than others, and these suffer more damage, particularly if they are of high density. Thus, the brain, the liver, and the spleen are more prone to be significantly damaged by a bullet wound than are the low-density and very elastic lungs, for example.

The permanent cavity is the hole left by the bullet. It is the track of tissue damage that occurs due to crushing and laceration as the bullet forces its way through the body. Handgun bullets tend to crush tissue and organs because of their large caliber. The volume of the permanent cavity correlates with bullet damage: bigger permanent cavities result in more damaging wounds.

To produce a large-volume permanent cavity, a bullet must penetrate deeply as well as be high caliber. This is why handgun wounds are often not as severe as rifle wounds: the combination of high caliber and low speed means that handgun bullets do not penetrate very far (but see fig. 7.7). Bears are hunted with rifles, not handguns, because a bullet needs to penetrate deeply in order to find the vital organs of a bear: a handgun bullet would just get it annoyed, which is not a good idea.

Fragmentation is a way of spreading the damage—increasing the per-

---

(1999), and Patrick (1989). Many Web sites discuss ballistic wounds and—a little out of our field—forensic ballistics.

Figure 7.7. A full-metal-jacket .45 cal automatic Colt pistol (ACP) round. Designed over a century ago as a "man-stopper," these rounds penetrate deeper than soft-nosed bullets (the jacket prevents mushrooming); this and the large caliber lead to a big permanent cavity wound. The ruler is marked in centimeters. Image from Wikipedia.

manent cavity and perhaps hitting a vital organ. Shotgun rounds fragment in the barrel and spread out during their trajectory; they do damage over a large area and can penetrate some distance at short ranges, causing massive damage. Other bullets are designed to fragment on impact, again doing significant damage. They dump their energy into tissue, instead of penetrating right through and taking a lot of energy with them, doing little damage unless they hit something vital. Soft-nosed and especially hollow-point bullets increase the size of the permanent cavity they create by deforming upon impact.[13] This effect is known as *mushrooming* because of the resultant bullet shape.

Mushrooming effectively increases the bullet caliber when it is inside the target. A low-caliber bullet can retain a lot of its muzzle velocity because it is less susceptible to aerodynamic drag over the course of its trajectory than is a high-caliber bullet. But if this low-caliber bullet is a hollow-point, then it mushrooms into a high-speed, high-caliber bullet inside a person, causing massive damage, as evidenced by the massive exit wound from such a round. Many hunting rifles fire bullets that deform in this way, but they are banned against humans in warfare by the Hague Convention of 1898. A low-power, small-caliber rifle used to hunt vermin benefits from hollow-point bullets; the low power means that little penetration occurs, but this is not so important for small prey. The mushrooming effect ensures a humane, quick kill. Other bullets are designed to re-

13. Expanding bullets are sometimes known as *dumdum* rounds, after the nineteenth-century British arsenal in Dum Dum, India, where they were developed.

sist deformation (e.g., full-metal-jacket military rounds). These penetrate deeper and are more effective against large game, such as elk or people.[14]

Another way that a bullet can increase its effective caliber, and thus increase the size of permanent cavity it creates (and so the damage it does), is by tumbling through tissue. If a bullet strikes bone, it may cause bone splinters to shoot out, acting like bullet fragments and doing secondary damage; but in addition, the bone may cause the bullet to tumble. As with mushrooming rounds, tumbling increases area and so helps to dump bullet energy into surrounding tissue, thus increasing the crushing and lacerations that are a feature of damaging wounds. Unlike bullets that mushroom, it is not illegal to use tumbling bullets against humans (after all, you can't expect the shooter to avoid hitting bones). The M16 rifle fires low-caliber rounds (.223) designed to tumble and produce large surface wounds.

High-velocity rounds striking tissue generate pressure waves (up to 200 atmospheres) and perhaps shock waves that propagate far from the wound track and can cause significant damage. For example, they can burst fluid-filled organs such as the heart, the vascular system, the bladder, and the bowel. (Such hydrostatic shock effects become dramatically credible if you see high-speed movie images of bullets striking soft fruit or full soft drink bottles.)[15] Neural damage also results from pressure waves, particularly from rifle rounds.

Howitzer How-To

For an artillery battery, terminal ballistics is about more than the effects of munitions on the victims. We have already seen two examples (MRSI and sound-and-flash) of artillery tactics used to locate targets; there are many

14. Wadcutters, bullets with a flat nose, look a little like hollow-point bullets, but they are not designed to damage animal tissue. Wadcutters are used in competition against paper targets; the flat nose seems to produce a clean, round hole with a less ragged edge, which helps with judging how close the shot came to the bull's-eye, for example.

15. These can be found on YouTube. See, for example, www.youtube.com/watch ?v=NQBsFoOVAeI. For a bullet passing through a block of ballistic gelatin, see www.youtube.com/watch?v=IL-liPFY5-I&feature=related. The complexity of high-speed fluid dynamics becomes quite evident in these films; thus, note how the target fluid explodes *toward* the bullet source as well as away from it.

Figure 7.8. Flechettes. These date from World War I and are 4 inches long. Modern flechettes are dispersed from beehive artillery rounds and are a quarter to a third of this length. Image from Wikipedia.

others, of course. Before even firing a salvo, however, the battery commander (you, let's say, just for the next few paragraphs) has to consider terminal ballistics because he needs to assess the target and choose an appropriate type of round that will achieve his object. The purpose of launching an artillery salvo against an enemy position is to achieve one of three levels of disablement—suppression, neutralization, or destruction. Suppression means that you are obliging the enemy to keep his head down: while you are firing at him, he is unable to function because he is too busy avoiding damage. Neutralization means he has suffered enough damage (10% casualties) to cause him to temporarily abandon operations after you have ceased firing. Destruction means just what it says: permanent loss of function (in practice this may mean 30% casualties).

These three levels represent an increasing intensity of fire and an increasing commitment of time and resources. How best to achieve the desired goal? Making this decision requires target assessment. Artillery targets are rarely uniform. You may be dealing with infantry out in the open or a line of trucks. The target may be a mixture of advancing armor and infantry, or a stationary group of buildings with dug-in guns and crews.

You make your assessment and choose appropriate ammunition. HE shells are very effective against infantry; high-explosive rocket-assisted (HERA) shells achieve the same results at a longer range; beehive rounds (fig. 7.8) are more effective if your targets are in the open.[16] Chemical

16. Beehive rounds contain many steel darts (8,000 for 155 mm howitzer ammunition) called *flechettes*, which spray out when the round is detonated mid-air by a time fuze. Flechettes are typically 1 inch long and break up when they strike a

rounds will choke or blister enemy personnel or cause damage to nervous systems; mines can be scattered from artillery rounds, denying ground to the enemy; smoke may mark his position or screen friendly positions from his view. Having chosen ammunition, you must decide the best method of attack. What aiming points? (Your targets may be distributed widely and unevenly.) What density and duration of fire? Having delivered your salvos you quickly move on ("shoot and scoot") to avoid counterfire, which nowadays can arrive very quickly, thanks to remote sensors such as radar. For this reason, the primary indirect fire support weapon today is the self-propelled howitzer (in the U.S. Army, the M-109 155 mm). If you are not mobile you must disperse and dig in, or else the terminal ballistics will be at your end of the shell trajectories.

Shell Wounds

In World War I more casualties were caused by artillery than by small arms, in marked contrast to the American Civil War, showing how artillery had come of age in the two generations that separated those conflicts. Artillery retains its dominance in the ethnic conflicts of today.[17]

Bullets are aimed at individuals; an artillery round is meant to kill many people. HE rounds cause casualties by blast and by fragmentation. The blast—the bang, flash, and flame—is local and fatal if you are caught in it. Casualties increase if the blast occurs in a confined space, say inside a tank or inside a building (due to structural collapse as well as the effect of concentrated blast pressure). The pressure wave can kill you without leaving a mark. Fragmentation spreads lethality further afield. The fragments may be contained within an ordnance round, or they may be parts of the exploding shell, as the projectile body is broken up by the HE blast. Fragments travel at high speed (around $1{,}000$ ms^{-1}) and are lethal to personnel, though they do only superficial damage to most hard targets. Thus, the "50% range" (the radius within which half the people are killed) for a 105 mm HE shell is roughly 50 feet (100 feet for a 155 mm shell). There is an optimum fragment size of 1.4 g (1/20 of an ounce). Any smaller and the fragments are slowed down too much by aerodynamic drag and so

---

target. Beehive rounds are so called because of the buzzing noise made by the flechettes.

17. War correspondent Anthony Loyd (1999) reported that the casualties in Grozny, Chechnya, resulting from the use of Russian heavy artillery were far worse than any of the horrors he had witnessed a few years earlier in Bosnia.

their lethal range is limited; any larger and the number of fragments is reduced so that casualties are reduced. As with many aspects of war, artillery casualties are statistical in nature—a fact of interest to munitions developers and military doctors but not, perhaps, to the casualties.

We have examined two aspects of terminal ballistics: armor penetration and wound damage. Toward the end of World War II, HEAT rounds fired from man-portable launchers were very effective at penetrating steel armor, because of the Munroe effect of shaped charges. Tank designers adapted by making armor thicker and more sloped. In the 1960s Chobham armor and reactive armor were invented. These techniques prevented, or perhaps only delayed, the eclipse of tanks as effective battlefield weapons. Missile designers countered by developing tandem-charge HEAT rounds and LRPs—and so the arms race between missiles and armor continues. Empirical formulas estimate the median speed required for an LRP or a bullet to penetrate steel armor.

Most bullets do not have enough momentum to knock down a person. Estimating the severity of a wound caused by a bullet from the bullet's characteristics is not easy, and the process of causing wound damage is not completely understood. Wound severity is known to increase with bullet caliber and speed, and is worse for expanding bullets. It correlates better with bullet momentum than with bullet kinetic energy. Pressure waves caused by bullets can do much damage at sites far from the bullet track.

Final Thoughts

I have taken you on an extended tour through the bang, whiz, and thud phases of a projectile's trajectory and, I think it fair to say, covered a wide range of projectile types. Ballistics encompasses every type of unguided projectile from a prehistoric sling stone to a HEAT missile, and from the oldest arrow to the latest anti-armor LRP. In between we have spent most of our time looking at small arms and bullets: their evolution through continuously (though unevenly) advancing technology, the ever-increasing complexity of their internal, external, and terminal ballistics.

We have taken two and a half thousand years to get from the projectile weapon of figure 1.1 to that of the illustration you see here. Only during

An M-4 carbine ejects a shell case during a U.S. Army firing exercise. The M-4 has selective fire options (semiautomatic or fully automatic) and many optional add-ons, including a high-tech optical sight and grenade launcher, as in this example. U.S. Department of Defense photo by Suzanne M. Day.

the last 10% of that period have we made significant progress in understanding projectile ballistics. Though not complete, this progress is impressive. As a scientist I feel bound to point out that advances in science, exemplified by ballistics, are measurable and undeniable, unlike advances in many other arenas of human civilization such as politics and diplomacy, the failure of which lead to the use of ballistics weapons.

Technical Notes

These technical notes establish results that are stated in the main text. I use math concisely; this approach packs a lot of analysis into each page. You may need a pencil and paper to get from one equation to the next, but the steps are straightforward, and the mathematically inclined reader will know how to unpack all of my analysis. Notes 1, 2, 4–8, 10–16, 21, and 22 are at high school or freshman undergraduate level, whereas notes 3, 9, and 17–20 require more advanced undergraduate math or physics. As you see, the later notes are generally more sophisticated. This reflects the more complicated ballistics that unfolds in the later chapters of the book; it also means that I can build on results established in earlier notes. There are few references to the technical literature here because most of this analysis is original.

NOTE 1. ARM THROW

Figure N1 shows a simple model of a throwing arm: two rigid rods of length L connected by a hinge. Both rods move with angular speed $\omega = \dot{\theta}$ about their axes of rotation—the fixed shoulder axis and the moving elbow axis. (I will use Newton's dot notation to indicate time derivatives, so that $d\theta/dt = \dot{\theta}$.) The hand end of this mechanical arm releases a rock so that the projectile is thrown along the y direction, as indicated. We should consider this action to be a sidearm throw (viewed from above) instead of an overarm throw to avoid the complicating effects of gravity.

The rock projectile is at location (x,y) where

$$x = L\cos(\theta) + L\sin(2\theta), \ y = L\sin(\theta) - L\cos(2\theta). \tag{N1.1}$$

The projectile speed along the x and y axes is thus

$$\dot{x} = -L\omega\sin(\theta) + 2L\omega\cos(2\theta), \ \dot{y} = L\omega\cos(\theta) + 2L\omega\sin(2\theta). \tag{N1.2}$$

To send the rock along the y direction we require $\dot{x} = 0$, which fixes the release angle (call it θ_0) at $\theta_0 = 36°$, so that the rock speed is $\dot{y} = 2.72L\omega$.

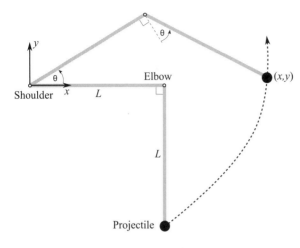

Figure N1. A simple hinged rod model of a throwing arm viewed from above. The origin of the coordinate system is at the shoulder end of this arm, so that the projectile is at location (x,y). Muscles rotate the upper arm relative to the shoulder, and the forearm relative to the upper arm, at the same rate $\omega = \dot{\theta}$.

Compare this speed with that of a stiff arm of the same length ($2L$) rotated at the same rate: $\dot{y} = 2L\omega$. So the whiplash action of the hinged arm adds 36% to projectile speed, in this model.

In fact the hinged arm can throw a little better than I have indicated. The maximum projectile speed along the y direction is found as the solution to the equation $d\dot{y}(\theta_0)/d\theta_0 = 0$, which yields $\theta_0 = 40°$ and $\dot{y} = 2.74L\omega$. To aim correctly (so that the projectile moves along the y-axis for this new θ_0) the arm should start at angle $\theta = -4°$ instead of $\theta = 0°$.

NOTE 2. JAVELIN LAUNCH ANGLE

Let us say that a javelin thrower runs up at a speed u and launches his javelin at angle θ to the horizontal at speed v (that is to say, the thrower's hand moves at speed v relative to his body). Let the horizontal and vertical axes be denoted x and z, respectively, so that the components of javelin velocity at launch are given by

$$\dot{x} = v\cos(\theta) + u, \quad \dot{z} = v\sin(\theta) - gt. \tag{N2.1}$$

Here, g is the acceleration due to gravity at the earth's surface. Integrating equations (N2.1) with respect to time, we obtain the horizontal distance traveled, and the javelin height, at a time t seconds after launch:

$$x = vt\cos(\theta) + ut, \quad z = vt\sin(\theta) - \frac{1}{2}gt^2. \tag{N2.2}$$

I am neglecting drag forces, to keep the analysis simple. If the javelin is thrown over level ground, it lands at time t_0, where $z(t_0) = 0$, so that $t_0 = (2v/g)sin(\theta)$. Substituting into equation (N2.2) we find the total horizontal distance traveled to be

$$R = x(t_0) = \frac{v}{g} (v \sin(2\theta) + 2u \sin(\theta)).$$ (N2.3)

Note that if $u = 0$ then we recover the expression usually given for the maximum range of a projectile, neglecting drag: $R = v^2/g$ for a projectile launched at angle $\theta = 45°$. For the case of a javelin thrower, however, we can see from equation (N2.3) that the optimum throwing angle is different from 45°: it occurs at angle θ_0, satisfying $dR/d\theta = 0$, which tells us that

$$\cos(\theta_0) = \frac{\sqrt{8v^2 + u^2} - u}{4v}.$$ (N2.4)

Equations (N2.3) and (N2.4) for maximum range and optimum launch angle are plotted in figure 1.2.

NOTE 3. STAFF SLING EFFICIENCY AND PROJECTILE SPEED

A model of the staff sling is shown in figure N3. For this model I am making some assumptions that simplify the calculations (though they will turn out to be pretty complicated anyway). I assume that the staff rotates at the lower end, as if the slinger was holding one end of the staff in one hand and rotating it with the other. This is a plausible slinger action, but surely not the best. A real staff slinger would move both hands. Also I assume in figure N3 that the staff is perfectly rigid: it does not flex like a fishing pole. Third, I assume that the slinger accelerates the staff angular rotation at a constant rate. The fourth simplifying assumption is that the sling has negligible mass. These four simplifications make the calculation tractable, but the answers that I get from this model will be only approximate. (Much of physics and engineering analysis involves compromises of this nature.)

I will use Lagrangian analysis to determine the projectile launch speed. The staff rotates with constant angular acceleration (call it α) so that $\theta = \frac{1}{2} \alpha t^2$. (The staff starts off horizontal, at time $t = 0$.) The gravitational potential energy of the staff plus sling is

$$V = \frac{1}{2} MgL \sin(\theta) + mg(L \sin (\theta) - l \cos(\phi)).$$ (N3.1)

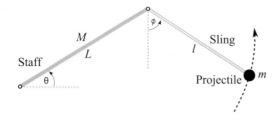

Figure N3. Staff sling model. Angles θ and ϕ define the orientation of staff and sling, respectively. The length and mass of the staff are L and M; those of the sling are l and m. The staff sling is swung in a vertical plane.

The kinetic energy of the staff plus sling system is as follows:[1]

$$T = \frac{1}{6} ML^2 \dot{\theta}^2 + \frac{1}{2}m\left[L^2\dot{\theta}^2 + l^2 \dot{\phi}^2 - 2Ll\dot{\theta}\dot{\phi}\sin(\theta - \phi) \right]. \qquad (N3.2)$$

The Lagrangian L is the difference between these two energies; the equation of motion for the staff sling is found from the Euler-Lagrange equation (one of the most useful equations in classical mechanics):

$$\frac{\partial L}{\partial \phi} = \frac{d}{dt}\left(\frac{\partial L}{\partial \dot{\phi}} \right), \text{ where } L = T - V. \qquad (N3.3)$$

Substituting from equations (N3.1) and (N3.2) gives us the equation of motion for the variable ϕ, the sling angle:

$$\ddot{\phi} = \frac{L}{l}\ddot{\theta}\sin(\theta - \phi) + \frac{L}{l}\dot{\theta}^2\cos(\theta - \phi) - \frac{g}{l}\sin(\phi). \qquad (N3.4)$$

Let us say that the slinger releases the projectile when it is moving at an angle of 45° to the horizontal, for maximum range. This angle corresponds to $\phi = 135°$ (see fig. N3). To see how ϕ changes over time from its initial value of 0°, I must resort to a computer to numerically integrate equation (N3.4). This calculation tells us the angular speed, $\dot{\phi}$, of the sling when the projectile is released, from which I can calculate projectile launch speed, v. The result, for different starting conditions, is plotted in figure 1.4.

The efficiency of the staff sling can be calculated from this simple model, as follows. Energy input to the staff sling by the slinger is $E = T + V$, with T and V given by equations (N3.2) and (N3.1), respectively. Projectile energy at launch is $E_p = \frac{1}{2} mv^2$ (plus a small amount of gravitational potential energy which

---

1. To establish equation (N3.2) you need to know the staff moment of inertia, which I have taken to be that of a uniform rod rotating about one end—i.e., $ML^2/3$. You will also need to determine the speed, v, of the projectile [v^2 is the term in square brackets in equation (N3.2)].

we can ignore here), and staff sling efficiency, ϵ is given by $\epsilon = E_p/E$. Efficiency is also plotted in figure 1.4.

NOTE 4. BURN RATE OF CORNED BLACK POWDER

Black powder burns from the outside in, and so burn rate is proportional to surface area, A. Consider a mass, M, of powder than has been corned into spherical grains of radius r. The volume of this powder is $V = M/\sigma$, where σ is black powder density. The number of grains is $N = V/v_g$, where $v_g = \frac{4}{3}\pi r^3$ is the volume of each grain. The total grain area is thus $A = 4\pi r^2 N = 3M/\sigma r$. Hence, burn rate is inversely proportional to grain radius.

NOTE 5. WINDAGE

This "back-of-the-envelope calculation" provides a rough ballpark figure for the windage of a very early gunpowder weapon such as a culverin or an arquebus. These crudely made weapons fired projectiles that were equally crude, so windage must have been significant.

We have a circuitous calculation to get through. First, note that the energy contained in the black powder charge is only partially converted into projectile energy; the rest is wasted. I will assume here that the three main contributions to wasted energy come from windage, friction, and heat. Other sources (such as sound energy) are considered negligible. So:

$$E = E_w + E_f + E_h. \tag{N5.1}$$

I estimate the windage energy as follows. Gas from the combustion of the black powder pushes the projectile out, but because of windage, some gas escapes without helping to propel the projectile. A crude estimate of the fraction of gas that usefully accelerates the ball is A_p/A_b, where $A_p = \pi r^2$ is projectile cross-sectional area (r is projectile radius) and $A_b = \pi(c/2)^2$ is the barrel cross-sectional area (c is caliber, or bore). So the energy lost to windage is roughly

$$E_w \approx E_0(1 - A_p/A_b) \approx \frac{2w}{c} E_0, \tag{N5.2}$$

where E_0 is the energy released by black powder combustion and w is windage (see fig. N5).

Now we need to estimate the energy lost to friction. Let us assume that the projectile hits the inside of the culverin or arquebus barrel n times as it accelerates along the barrel length. It is reasonable to expect that the projectile will rattle around as it moves along the barrel because it is only loosely

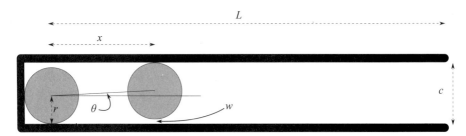

Figure N5. Windage, w, is the difference $c - 2r$ between caliber and ball diameter. This difference permits the ball to move at an angle θ to the barrel axis. The ball will strike the barrel n times, where n is the largest whole number that is less than L/x.

fitted and because neither projectile nor bore are particularly smooth, given the technological limitations in manufacturing. Each time the projectile hits the barrel it loses a little energy. Say E_p is the projectile energy just prior to a collision; then the energy afterwards is eE_p, where e is the "coefficient of restitution." So if the ball comes away with 90% of the energy it had before the collision, then $e = 0.90$. Thus the energy lost due to friction is

$$E_f \approx n(1 - e)E_p \approx n(1 - e)\epsilon E_0. \tag{N5.3}$$

In equation (N5.3) ϵ is the efficiency of our crude firearm; it is the fraction of black powder energy that is converted into projectile energy. How do we estimate n? From figure N5 it is easy to see that $n \approx \theta L/w$, where L is barrel length and θ is the (small) angle of the projectile within the barrel. In a perfect world, θ would be zero, and so the projectile would not strike the sides of the barrel at all, but the manufacturing imperfections of these early weapons means that we must expect that they shoot off to the side a little.

Putting together equations (N5.1–N5.3):

$$E \approx E_0 \left(\frac{2w}{c} + \epsilon(1 - e)\frac{L}{w}\theta + \rho \right). \tag{N5.4}$$

I have expressed the heat energy as a fraction, ρ, of the black powder energy: $E_h = \rho E_0$.

You may well be wondering what energy considerations have to do with estimating windage. Here is the payoff. We can assume that over the first couple of centuries gun manufacturers tinkered with their designs to get the best results. In particular, they hit upon the windage that wasted the least black powder energy, all other variables being the same. I can calculate this minimum wasted energy from equation (N5.4) by minimizing E with respect to w. The result of this calculation is an optimum windage of

$$w_{opt} = \sqrt{\frac{1}{2}\,\epsilon(1-e)\theta cL} \qquad\qquad\qquad\qquad (N5.5)$$

and a minimum energy loss of

$$E_{min} = 4w_{opt}E_0/c + \rho E_0. \qquad\qquad\qquad\qquad (N5.6)$$

Note also that $E_{min} \approx (1-\epsilon)E_0$ and so, finally,

$$w_{opt} \approx \frac{1}{4}\,(1-\epsilon-\rho)c. \qquad\qquad\qquad\qquad (N5.7)$$

My back-of-the-envelope calculation has led to equation (N5.7) for optimum windage in terms of gun efficiency and caliber. The earliest figures I have obtained are from the nineteenth century; these report efficiency measurements of about 30% ($\epsilon = 0.3$) and heat loss of 50% ($\rho = 0.5$).[2] Earlier weapons would have been less efficient—say $\epsilon = 0.2$.

I will adopt these values for the earlier weapon that we are considering. The prediction we obtain will be approximate, but in the right ball park. For these values, equation (N5.7) gives a windage of $\frac{1}{13}$ of the caliber (7.5%). This figure compares with $\frac{1}{16}$ of a caliber (6.25%) for a matchlock from the English Civil War period (mid-seventeenth century), and $\frac{1}{40}$ of a caliber (2.5%) for the windage of a better-engineered nineteenth-century weapon. The somewhat larger value that I have obtained reflects the crude engineering of the earlier guns.

My calculation cannot be applied to better-engineered muzzle-loading weapons because, for these, n is a small number (such as 1 or 2). In this case the second term of equation (N5.4) is a constant and so the optimum value for windage is predicted to be zero, as is easily seen. The reasonable value I have found for early muskets serves to show how the various parameters of early firearm internal ballistics interact—they are not independent—and how we can apply back-of-the-envelope physics to provide ballpark estimates.

NOTE 6. SCALE EFFECTS

In figure N6a I sketch the cross-section of an early firearm or cannon with a ball projectile. The mass of the projectile is denoted by m and that of the gunpowder by m_g. Barrel dimensions are as indicated. Suppose that we want to increase the caliber of this weapon; how do the other parameters scale? The following simple calculation provides an estimate.

2. See Hatcher (1962). The efficiency of modern weapons is about the same—see Klingenberg (2004)—although the muzzle velocities are greater.

(a)

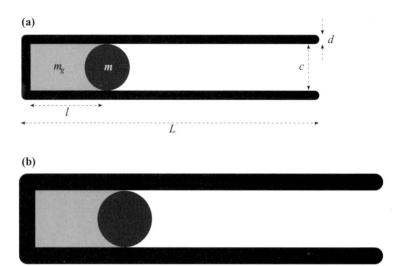

(b)

Figure N6. Scale effects for a black powder gun. (a) Gun cross-section: L = barrel length, l = barrel length taken up by powder, c = caliber, d = barrel thickness, m = projectile mass, m_g = powder mass. (b) The gun is scaled up by 20% (i.e., λ = 1.2). Note the barrel thickness.

Say the caliber is to increase by a factor λ, to $c' = \lambda c$ (so, for example, if the caliber is increased by 20% then $\lambda = 1.2$). It clearly follows that the bore area must increase to $A' = \lambda^2 A$ (where $A = \pi c^2/4$) and the projectile mass must increase to $m' = \lambda^3 m$. For these old guns, the mass of powder was generally set at a fixed fraction of the projectile mass, and so $m'_g = \lambda^3 m_g$. Powder mass is proportional to powder volume (assuming that packing density is unchanged), which means that $l' = \lambda l$.

We need to know how peak pressure scales. (Peak pressure refers to the maximum pressure inside the barrel due to burning powder.) Say $P' = \lambda^p P$ where the exponent p is to be determined. The force of expanding gas pushing against the projectile is $F = PA$, and so this force scales as $F' = \lambda^{p+2}F$. The projectile acceleration, a, is determined from Newton's second law, $F = ma$ which leads to $a' = \lambda^{p-1}a$. Assume that the scaled-up gun is to produce the same acceleration as the original; this requires $p = 1$. Thus, pressure must scale as $P' = \lambda P$ and force as $F' = \lambda^3 F$. This force pushes the projectile out of the barrel; by Newton's third law it also pushes against the closed end of the barrel, causing stress. If the "safety factor"—the ratio of barrel strength to stress—is unchanged by scaling, then $S'/F' = S/F$. I will assume that the strength, S, of the barrel is proportional to barrel thickness, d, which leads to $d' = \lambda^3 d$. If $\lambda = 1.2$ then the scaled-up barrel looks like figure N6b.

The main result obtained here is that barrel thickness increases faster than caliber.[3] I will leave it as an exercise for the interested reader, as the textbooks say, to show that barrel mass increases even faster. Larger guns are disproportionately heavier than smaller ones.

NOTE 7. BARREL LENGTH VS. MUZZLE SPEED

Figure N7 shows a ball being fired from a musket or from a smooth-bore cannon. The ball speed when at position x (at time t) is denoted $\dot{x}(t)$. When the ball reaches the muzzle (at time t_m), its speed is v, so that $\dot{x}(t_m) = v$. Other gun parameters are shown in the figure. Here I will make use of a simplified mathematical model of musket internal ballistics to demonstrate that muzzle speed, v, depends upon barrel length, L.

The first simplifying assumption is that the black powder charge burns at a constant rate, in which case the mass of powder remaining unburned at a time t after ignition is given by

$$m_{gunpowder}(t) = m_g - \dot{m}_g t. \tag{N7.1}$$

In equation (N7.1) the initial mass of powder is m_g and the burn rate is $\dot{m}_g$, so that the burn time is

$$t_b = m_g / \dot{m}_g. \tag{N7.2}$$

If a fraction, f, of the powder mass is turned into gas, then the mass of gas at time t after powder ignition is

$$\begin{aligned} m_{gas}(t) &= f\dot{m}_g t, \quad 0 < t < t_b \\ &= fm_g, \qquad t > t_b. \end{aligned} \tag{N7.3}$$

I will assume that the gas expands isothermally according to the ideal gas law, so that

$$PV = km_{gas}(t). \tag{N7.4}$$

3. You might question some of the assumptions that I made for this scaling calculation. For example, perhaps barrel strength is proportional to barrel cross-sectional area (taken perpendicular to the barrel axis) instead of being proportional to thickness. If so, then the scaling behavior is more complicated, in general; if caliber is much greater than thickness ($c \gg d$), then the scaling behavior for this alternative assumption simplifies to $d' = \lambda^2 d$ (so barrel thickness still increases more rapidly than does caliber).

(a)

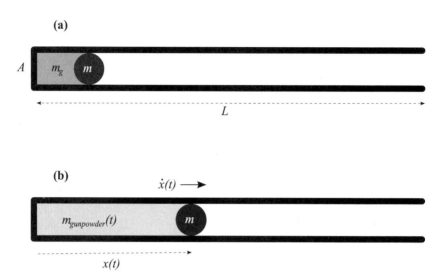

(b)

Figure N7. Internal ballistics of a black powder gun. (a) Initial setup: A = bore area, L = barrel length, m = projectile mass, m_g = initial powder mass. (b) Situation at a later time, t: $x(t)$ = projectile position, $\dot{x}(t)$ = projectile velocity, m = projectile mass, $m_{gunpowder}(t)$ = mass of remaining gunpowder.

Equation (N7.4) is a major simplification, because the gas expansion is certainly not isothermal.[4] We know that the gas heats up considerably as it burns and expands. I make this assumption to render the calculation tractable, but the price paid is that we can regard the calculation results as only approximate. In equation (N7.4) P is the gas pressure and V is gas volume; these can be written as follows (see fig. N7):

$$P = \frac{F}{A} = \frac{m\ddot{x}(t)}{A} , V = Ax(t). \tag{N7.5}$$

The parameter k of equation (N7.4) depends upon several factors, such as the black powder packing density and the gas temperature. I am assuming constant temperature, and so k is a constant throughout the internal ballistics

4. An isothermal process is one in which temperature does not change. So, in this case, by assuming the musket's internal ballistics to be an isothermal process, I am assuming that the gas does not heat up as it expands and burns. An alternative assumption is that the propellant gas is an ideal gas that expands adiabatically, meaning that no heat is transferred from the burning powder to the gas or to the gun barrel. We will see later how this different assumption leads to slightly different predictions.

process, from $t = 0$ (ignition) to $t = t_m$ (ball exiting barrel). Combining equations (N7.3)–(N7.5) yields the following equation of motion for the musket ball:

$$x(t)\ddot{x}(t) = \alpha t, \quad 0 < t < t_b \quad \text{where } \alpha = \frac{kf\dot{m}_g}{m}. \qquad \text{(N7.6)}$$
$$= \alpha t_b, \quad t > t_b,$$

These differential equations are not difficult to solve. There are two cases of interest to us. First, for an early musket or arquebus the powder burns quickly —in this case the burn time is much shorter than the time it takes the ball to exit the barrel (i.e., $t_b \ll t_m$). The second of equations (N7.6) applies, and the solution is

$$v = \sqrt{\alpha t_b \left(3 + 2\ln\left(\frac{L}{l}\right)\right)}. \qquad \text{(N7.7)}$$

Equation (N7.7) shows that, for a fast-burning powder, muzzle speed increases very slightly as barrel length increases (assuming the same powder charge for all barrel lengths). For example, if L doubles from $10l$ to $20l$ then muzzle speed increases by only 9%.

The second case of interest is that for which $t_b = t_m$ (the last of the powder is burned just as the ball exits the muzzle), in which case the first of equations (N7.6) applies; its solution can be expressed as

$$\dot{x}(t_b) = v = \sqrt{3\alpha t_b}, \quad x(t_b) = L = \sqrt{\frac{4}{3}\alpha t_b^3}. \qquad \text{(N7.8)}$$

Eliminating t_b yields

$$v = \left(\frac{9}{2}\alpha\right)^{1/3} L^{1/3}. \qquad \text{(N7.9)}$$

This prediction—that muzzle speed increases as the cube root of barrel length —is close to empirical observations made in the nineteenth century for black powder artillery that muzzle speed increases as the fourth root of barrel length. [See chapter 2, and in particular fig. 2.8 where equation (N7.9) is plotted against real data.] The agreement is approximate—hardly surprising given the rough nature of this calculation—but close enough to show that we are on the right track.

I can change the isothermal assumption of equation (N7.4) into an adiabatic assumption: no heat is lost from the expanding gas. It can change temperature but not lose heat to the barrel, for example. Again, this is not exactly true; it is a different type of approximation. For adiabatic expansion, equation (N7.4) is replaced by $PV^\gamma = km_{gas}(t)$ where the exponent γ is approximately equal to $\frac{7}{5}$. Repeating the analysis leads to the prediction that muzzle speed increases as the fifth root of barrel length. So, these two simple approximations lead to predictions that neatly frame the observed relationship between speed and barrel length.

Equations (N7.8) can be expressed in a different form:

$$v = \sqrt{\frac{3kfm_g}{m}}, \quad t_m = L\sqrt{\frac{3}{4}\,m/kfm_g}. \tag{N7.10}$$

These equations are applied in chapter 2 to several nineteenth-century firearms; the results are, once more, close enough to the real world to give confidence in the calculations of this note.

It is not difficult to extend the calculation to a third case, that in which $t_b > t_m$. This case is suboptimal: the projectile is expelled from the barrel before all the powder is burned. Consequently, some of the powder is wasted. This case is discussed in chapter 2; it explains some seemingly odd test results which showed that very large charges actually reduced muzzle speed.

Finally, let us calculate the work done in moving the musket ball or cannonball. From elementary thermodynamics we have the following expression for work done: $W = \int dV\,P$, where the integral covers the volume of the gun barrel. Substituting for pressure and volume from equation (N7.5), the integral is readily evaluated to yield $W = \frac{1}{2}mv^2$. This is just the kinetic energy of the ball, which tells us that none of the energy of the black powder has been wasted—all has been used to accelerate the ball. This observation reflects another simplifying (and unrealistic) assumption of my calculation: no friction between barrel and ball. In reality, there is significant friction, the effects of which are discussed in the next note.

NOTE 8. EFFECTS OF FRICTION

Friction between the projectile and the barrel influences internal ballistics. We can understand this most clearly for the case of weapons that use fine (uncorned) black powder, which burns quickly. From the second of equation (N7.3), which applies for such weapons, and from equations (N7.4)–(N7.5), it is easy to show that the force acting on a projectile within the barrel at position x is given by $F = kfm_g/x$. Now let's add a friction force, F_{fr}, and assume that it is the same at all points along the barrel:

$$F = kfm_g/x - F_{fr}. \tag{N8.1}$$

This equation tells us that, for long barrels, the force can become negative. More specifically, if the barrel is long enough so that x can exceed kfm_g/F_{fr}, friction slows down the projectile before it exits the barrel, as Charles V observed (see chapter 2). If the barrel length increases further, then friction acts for a longer time and reduces muzzle speed more. Optimum barrel length occurs for $F = 0$:

$$L_{opt} = \frac{kfm_g}{F_{fr}} .$$

(N8.2)

Note from equation (N8.2) that the optimum barrel length increases with charge, but this equation is valid only when all the powder burns before the projectile leaves the barrel.

NOTE 9. RIFLING

With rifling, helical grooves are cut out of the gun barrel, leaving raised "lands" that are in contact with the bullet as it travels down the barrel. A recent development is polygonal rifling; in this case, the interior barrel cross-section is a polygon which twists down the barrel. This design is reckoned to reduce barrel wear and so increase firearm service life.

Twist is measured as the distance (here denoted d_t) corresponding to one complete turn of the rifling grooves, so a 10-inch twist means the rifling makes one complete rotation around the barrel interior over 10 inches of barrel length. "Gain twist" increases the twist rate near the barrel mouth; this may cause less damage to a bullet by allowing for the angular acceleration as the bullet progresses along the barrel. Twist rates are higher for long rifle bullets than for short handgun bullets; they are higher still for artillery shells. This is because the aerodynamic stabilization conferred by a spinning projectile must be stronger for longer projectile lengths, to counter the stronger destabilizing forces. We learn why in chapter 5.

The Greenhill formula is a long-standing rule of thumb which tells us how much twist a given bullet should be given, to provide for a stable trajectory:[5]

$$d_t = \alpha \frac{c^2}{b} .$$

(N9.1)

The constant α is 150 in most cases (rising to 180 for very-high-speed projectiles). The bullet length is b, and c is caliber, or bullet diameter. We can use this formula to estimate the fraction of projectile energy taken up by rotation. This fraction is

$$f_{rot} = \frac{1/2 \ I\omega^2}{1/2 \ I\omega^2 + 1/2 \ mv^2} .$$

(N9.2)

5. The formula is named for George Greenhill, a professor of mathematics at Woolwich Artillery College, England, in the late nineteenth century.

In equation (N9.2) m is the bullet mass, v is bullet muzzle speed, ω is bullet angular speed, and I is the moment of inertia for a bullet rotating about its longitudinal axis. The moment of inertia, I, can be written approximately as $I \approx 1/8\ mc^2$, and angular speed is $\omega = 2\pi v/d_t$. Putting all this together,

$$f_{rot} \approx \frac{b^2}{s + b^2}\ , \text{where } s = 2\left(\frac{\alpha c}{\pi}\right)^2. \tag{N9.3}$$

For example, a 30-caliber bullet that is 1 inch long (c = 7.62 mm, b = 25.4 mm), fired with a muzzle speed of 2,000 ft/s (606 m/s), will require a twist of d_t = 13½ in (0.34 m) and will spin at 1,780 Hz. From equation (N9.3) we see that spinning takes up about 0.24% of projectile energy. This is typical for a small arms bullet: almost all the energy is due to speed, not spinning.

NOTE 10. GRAIN SHAPE AND BURN RATE

The volume of a solid cylindrical grain of powder is

$$V = \pi R^2 L, \tag{N10.1}$$

where R is cylinder radius and L is cylinder length. The surface area of the grain, at a time t after it starts to burn, is

$$A(t) = 2\pi r(t)L, \tag{N10.2}$$

where $r(t)$ is the radius at time t. I am assuming that R is much less than L (so that the grain is long and thin), which means that I can neglect the shortening of the grain as the ends burn. For a constant burn rate $\dot{r}$, the radius at time t is simply

$$r(t) = R - \dot{r}t. \tag{N10.3}$$

The burn time for this grain is $t_b = R/\dot{r}$. The volume of gas generated by the burning powder grain is proportional to powder volume. Say the constant of proportionality is α (where $\alpha \approx 1,000$ for smokeless powder). Thus, the change dV_g in gas volume caused by a change dr in grain radius is

$$dV_g = -\alpha A dr. \tag{N10.4}$$

Substituting from equation (N10.2) and integrating yields $V_g(r) = \pi \alpha L\ (R^2 - r^2)$. Substituting from equation (N10.3) gives us, finally,

$$V_g(t) = \pi \alpha L\dot{r}t(2R - \dot{r}t). \tag{N10.5}$$

[Note that $V_g(t_b) = \alpha V$, as it should.] Equation (N10.5) tells us how fast that gas is generated by a solid grain of burning powder. This is the so-called "degressive" burn discussed in chapter 3.

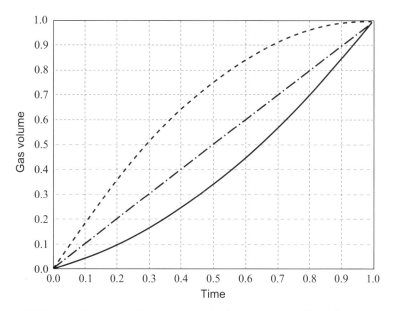

Figure N10. Propellant gas volume vs. time for burns produced by different powder grain shapes: degressive (dashed line), neutral (dash-dot line), and progressive (unbroken line).

The equivalent expression for a "progressive" burn is found similarly. Assume there are n circular holes through the length of the grain, as in figure 3.6, so that the grain volume is

$$V = \pi R^2 L - n\pi r_1^2 L. \tag{N10.6}$$

where r_1 is the initial radius of the internal holes. The area burning at time t is

$$A(t) = 2\pi(R + nr_1 + (n-1)r(t))L, \text{ where } r(t) = \dot{r}t. \tag{N10.7}$$

Proceeding as before leads to the following expression for gas volume:

$$V_g(t) = 2\pi\alpha L \left[(R + nr_1)\dot{r}t + \frac{1}{2}(n-1)(\dot{r}t)^2 \right]. \tag{N10.8}$$

Equation (N10.8) expresses the rate at which a multiperforated grain generates gas when it burns. Burn time cannot be easily defined for this case because the holes change shape after they begin to coalesce, when their radii become large enough (see fig. 3.6). For $n = 1$ the progressive burn of equation (N10.8) reduces to a neutral burn. Equations (N10.5), and equation (N10.8) with $n = 1$ and $n = 7$, are plotted in figure N10. The graphs are scaled so that the grains generate the same volume of gas in the same time. For real grains, this would be achieved by adjusting the grain sizes.

NOTE 11. BURN RATE AND BARREL LENGTH FOR SMOKELESS POWDER WEAPONS

Figure N11 shows a bullet traveling down the barrel of a modern rifle or a shell inside an artillery piece. Assume that the pressure varies with the projectile's position, x, in the barrel as follows:

$$P(x) = P_0 ax\, exp(1 - ax). \tag{N11.1}$$

This distribution produces the curves shown in figure 3.7a: the pressure peaks at different positions along the barrel, depending upon the value assigned to the parameter a, but the value of peak pressure is the same in all cases. The question we address here is this: Is there an optimum choice for pressure distribution curve, and if so, how does it depend upon barrel length?

First, note that we are not truly modeling the internal ballistics of a modern firearm here because we are not saying how the pressure distribution curve arises. We know that expanding gas generates pressure (and increases temperature, causing the pressure to rise further), and we know that the pressure is influenced by projectile position within the barrel. Here we do not attempt to model this complicated process (much more complex than the internal ballistics of black powder weapons, modeled in technical note 7), but simply parameterize the pressure curve and ask if there is an optimum shape.

We will allow for friction, so that the force acting upon the projectile of figure N11 is given by

$$m\ddot{x} = P(x)A - f, \tag{N11.2}$$

where $\ddot{m}$ is projectile acceleration (dots indicate time derivatives), A is barrel or projectile cross-sectional area, and f is the force of friction between projectile and barrel (assumed constant).

Equation (N11.2) integrates to yield projectile velocity. At the muzzle this velocity is found to be

$$v(a,L) = \sqrt{\frac{2}{m} \left[\frac{eP_0A}{a}(1 - (1 + aL)exp(-aL)) - fL \right]}, \tag{N11.3}$$

where $e = 2.71828$ is the base of natural logarithms. So, muzzle speed depends upon the pressure curve parameter a and barrel length L. The optimum choice for barrel length satisfies the equation $dv(a,L)/dL = 0$, which leads to

$$aL exp(-aL) = \frac{f}{eP_0A}. \tag{N11.4}$$

(This choice also corresponds to zero force acting on the projectile at $x = L$, as well as to maximum velocity.) Equation (N11.4) tells us the barrel length that

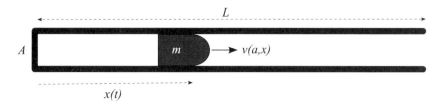

Figure N11. A bullet or artillery round traveling down a barrel of cross-sectional area A and length L. At time t the projectile position is $x(t)$. The projectile velocity at position x is $v(a,x)$, so that the muzzle velocity is $v(a,L)$. Here, a is a parameter that describes the propellant gas pressure curve.

must be chosen to maximize $v(a,L)$. But what is the best choice for the remaining parameter a? Substituting equation (N11.4) into (N11.3), eliminating f, leads to $v(a)$, which is plotted in figure 3.7b. Note that there is an optimum choice: $a \approx 5/L$. Also in figure 3.7c I plot the firearm efficiency; this is the fraction of gunpowder energy that is taken away by the projectile: $\epsilon = \frac{1}{2}mv^2/W$, where

$$W = A \int_0^L dx\, P(x) \tag{N11.5}$$

is the released gunpowder energy. The fraction of energy used to overcome barrel friction, fL/W, is also plotted in figure 3.7c. For the optimum choice of a and L, efficiency is 82%. The remaining 18% of energy is used to overcome friction.

You might look at the pressure curves of figure 3.7a and wonder why the choice $aL = 5$ is optimum if, as suggested by equation (N11.5), imparted gunpowder energy is the area under the curve. Surely, larger energy (bigger area) will produce faster projectiles. The largest area occurs for $aL = 1.8$, so why isn't this choice optimum? Because of bullet dynamics. Projectile inertia and the force of friction, as well as the pressure curve, influence projectile speed, as we see in equation (N11.3).

We can show that this simple pressure-curve model produces reasonable numbers by comparing these numbers with real data. An M16 rifle firing NATO 5.56×45 mm rounds produces a peak gas pressure of between 43 kpsi and 62 kpsi. Let us pick a middle value of 53 kpsi (3.65×10^8 N/m², or 3,600 atm). The barrel length is 20 inches (0.5 m). Bullet mass is 4 g and cross-sectional area is 2.4×10^{-5} m². Equation (N11.3) predicts that the optimum muzzle speed for this combination of bullet and rifle is 3,213 ft/s (974 m/s).

In practice the muzzle speed is in the range 3,050–3,350 ft/s, so the model is making sensible predictions.[6]

NOTE 12. MULTIPLE ROUND SIMULTANEOUS IMPACT (MRSI)

In figure N12 a howitzer is firing over a hill at a target. For the target to be hit with five rounds simultaneously, the projectile muzzle speeds and trajectories need to be different. Let us say the gun is capable of ejecting a spent case, loading a new round, and re-aiming, all within a short time interval, $\tau = 1$ s. The five shots of figure N12 are given a label, n, with $n = 0$ corresponding to the first shot fired and $n = 4$ to the last. Neglecting aerodynamic drag, the target range, R, is expressed in terms of muzzle speed, v_n, and launch angle, a_n:

$$R = \frac{v_n^2}{g} \sin(2a_n). \tag{N12.1}$$

Here I will show you that it is possible, with realistic intervals, τ, for the howitzer to fire several shots in succession, all of which arrive at the target at the same time.

The first shot fired takes a time

$$t_0 = \sqrt{\frac{2R}{g} \tan(a_0)} \tag{N12.2}$$

to reach the target. For example, if $R = 10$ km and $a_0 = 20°$, then $t_0 = 27.2$ s. We require that the second shot takes a shorter time to reach the target: $t_1 = t_0 - \tau$ (because the second shot is fired τ seconds after the first). In general, if

$$t_n = t_0 - n\tau, \tag{N12.3}$$

then all the shots will hit the target simultaneously. If the launch angles are related by

$$a_n = a_0 - nb, \tag{N12.4}$$

then equations (N12.2)–(N12.4) tell us the angular increment (after some calculation):

$$b = \alpha \sqrt{\frac{g}{2R}} \frac{\sin(2a_0)}{\sqrt{\tan(a_0)}}. \tag{N12.5}$$

6. M16 data is from two Wikipedia articles: "M16 Rifle" and "5.56×45 mm NATO."

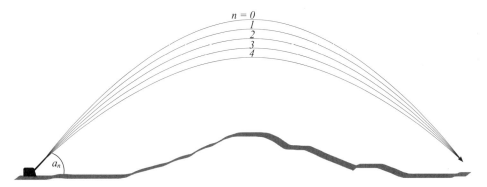

Figure N12. Multiple round simultaneous impact (MRSI): Five artillery rounds fired in quick succession at different elevation angles strike the same target area simultaneously.

Equation (N12.5) is valid for small values of b. So we have calculated the required decrease in elevation angle of the gun barrel, after every shot. For the same parameters as before ($\tau = 1$ s, $R = 10$ km, and $a_0 = 20°$) we find that $b = 1.35°$.

Inverting equation (N12.1),

$$v_n = \sqrt{\frac{gR}{\sin(2a_n)}} \, , \tag{N12.6}$$

and substituting from equations (N12.4) and (N12.5), we find

$$v_n = V_0 + nu, \text{ where } u = g\tau \, \frac{\cos(2a_0)}{2 \sin(a_0)} \, . \tag{N12.7}$$

Again, in calculating v_n I have assumed that b is small. For the same parameter values as before, we find $v_0 = 390$ m/s and the muzzle speed increment is $u = 11.0$ m/s.

The problem is solved. We see that the howitzer barrel must be lowered by an angle b between shots and the muzzle speed increased by u. This is feasible: nowadays it is possible to control the amount of powder in a round (made easier for howitzers because the charge is separate from the projectile) and hence the muzzle speed, while computer-driven fire-control mechanisms take care of the aiming. In reality, the calculations are more complex than indicated here because, of course, aerodynamic drag influences flight times and greatly complicates the mathematics. (I will investigate drag in a later note.) Nevertheless, my simplified calculation serves to show that MRSI is achievable.

NOTE 13. RECOIL

According to Newton's third law, the momentum of projectile and propellant gas must produce a recoil:

$$Mv' = m_{gas}v_{gas} + mv = kmv. \tag{N13.1}$$

In equation (N13.1) Mv' is recoil momentum; where M is the mass of the gun, rifle or pistol; and v' is the recoil velocity (in the opposite direction to projectile velocity). Propellant gas mass, m_{gas}, and velocity, v_{gas}, are variable (different gas molecules move at different velocities); thus, equation (N13.1) represents an average. I will assume that the effect of the gas is to raise the recoil above that of projectile momentum, mv, by a factor k, so that $k > 1$. Because the projectile momentum for most guns will exceed the gas momentum, k will be not much greater than 1.

Projectile energy and recoil energy are written as

$$E_{proj} = \frac{1}{2} mv^2, \quad E_{recoil} = \frac{1}{2} Mv'^2. \tag{N13.2}$$

Substituting for v' from equation (N13.1) yields the following expression for the ratio of recoil energy to projectile energy:

$$\frac{E_{recoil}}{E_{proj}} = k^2 \frac{m}{M}. \tag{N13.3}$$

This ratio is small for handguns and most rifles. For example, a 2-pound handgun with a 115-gr bullet and 5 gr of powder (corresponding to masses of 0.91 kg, 0.0075 kg, and 0.00032 kg, respectively) will have $k \approx 1$ and a recoil energy that is less than 1% of bullet energy. A low-caliber rifle with a 40-gr charge and a 40-gr bullet will have a larger k value but also greater weight, so the energy ratio will still be less than 1%. Note that, from equation (N13.3), a heavy weapon gives up less energy to recoil than does a light weapon.

A similar calculation shows that muzzle speed (relative to the ground) increases if the firearm does not recoil—for example, if a rifleman is braced against a tree or if an artillery piece is dug into the ground.[7] In such cases the effective muzzle speed increases by a factor of $\sqrt{1 + k^2 m/M}$.

To see how recoil can adversely affect handgun aim, consider figure N13. Recoil causes the barrel to kick upward—in other words, to rotate about an axis located on the pistol grip, here denoted by X. A back-of-the-envelope

---

7. See, e.g., Weinstock (2002) for the increased range of a cannon that is braced against a tree.

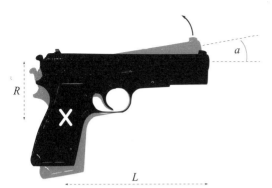

Figure N13. Handgun recoil can lead to the phenomenon of *muzzle rise,* or rotation, as well as to a backward kick. This happens because the recoil momentum is above the center of rotation (**X**). We can estimate the angle, *a,* of muzzle rise.

calculation will show how much the kick influences aim. The bullet is inside the barrel for a time of, roughly, $\tau \sim L/v$, where v is muzzle speed. Recoil causes the pistol barrel to rotate at a rate $\dot{a} = v'/R$, where R is defined in figure N13 and v' is the recoil velocity. So the recoil angle is $a \sim v'\tau/R$. Substituting for τ and, from equation (N13.1), for v' we find that

$$a \sim \frac{mL}{MR} \tag{N13.4}$$

for a pistol with $k \approx 1$. (A more sophisticated calculation, which takes into account the changing barrel angle as the bullet travels down it, yields the same angular deviation.) For example, a bullet with a mass of 3 g fired from a pistol with a mass of 1 kg, a barrel length of 15 cm, and $R = 5$ cm will kick at an angle $a \sim 0.5°$, so that the bullet trajectory will be half a degree above the aim point. This is enough to miss a target at 75 feet distance by 6 feet. So, pistol shooters, hold the grip firmly.

A final recoil calculation concerns muzzle brakes. If the pistol of figure N13 had a muzzle brake (it does not) that directed propellant gas vertically upward, then a simple torque calculation shows that a mass $m_{gas} \sim mR/L$ of such gas would be enough to counter the kick: there would be no rotation (though there would still be a linear recoil movement of the pistol).

NOTE 14. TRANSITIONAL REGION

How far from the barrel does the region of transitional ballistics extend? We can roughly estimate this distance as follows. From figure N14 you can see that I am assuming that the blast of gas takes a hemispherical shape. This assumption is supported by more detailed calculations and by observing photos of

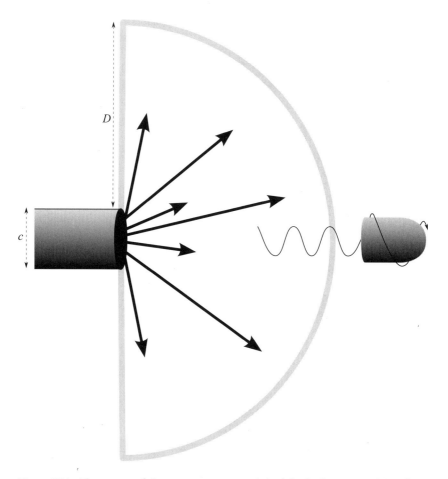

Figure N14. The extent of the transition region (*D*) of the high-pressure blast from a barrel of caliber *c* can be roughly estimated from simple geometrical arguments, outlined in the text.

muzzle flash.[8] Of course, this assumption is an approximation, but it is one that is good enough to provide us with a rough estimate. Gas pressure is well above atmospheric levels when the projectile emerges from the barrel. For the optimum pressure curve of figure 3.7a (corresponding to $aL = 5$) you can see that gas pressure at the muzzle is about 10% of the peak pressure. For a NATO 5.56 mm round fired from an M16 rifle (the case that we were considering in technical note 11), peak pressure was 53 kpsi (3,600 atm), and so I assume

8. You might argue for a spherical shape instead of hemispherical, but this makes little difference to my calculation.

that the pressure at the muzzle is 5.3 kpsi or 360 atm. The key assumption for this calculation is that transitional ballistics applies for gas pressures that exceed atmospheric pressure; for lower gas pressures the projectile is not affected by this gas—it is much more influenced by aerodynamic drag.

In figure N14 the hemisphere of gas spread out to distance D from the muzzle has volume

$$V_{gas} = \frac{2}{3}\,\pi D^3.$$
(N14.1)

Pressure is diluted as the blast extends beyond the muzzle. It decreases to atmospheric pressure when the following equation is satisfied:

$$P_m V_b = P_{amb}(V_b + V_{gas}).$$
(N14.2)

Here P_m is gas pressure at the muzzle, and P_{amb} is ambient (atmospheric) pressure. Barrel volume, V_b, is given by

$$V_b = \pi \left(\frac{1}{2}c\right)^2 L,$$
(N14.3)

where c is caliber and L is barrel length.

From equations (N14.1)–(N14.3) the extent of the transition region is estimated as

$$D \approx \left(\frac{3c^2 L}{8}\,\frac{P_m}{P_{amb}}\right)^{1/3}.$$
(N14.4)

Substituting the M-16 data from technical note 11 yields $D_{M16} \approx 0.13$ m (about 5 in). For the larger guns of the USS *Iowa* (fig. 4.1) a similar calculation (assuming the same value for muzzle pressure) yields $D_{IOWA} \approx 10$ m (33 ft).

For the *Iowa*, we can estimate the effect of the pressure blast upon the water surface. The pressure required to depress the water level by 1 foot is about $\frac{1}{34}$ P_{amb}, and so the distance out from the muzzle that water level is depressed by 1 foot (on average) is $34^{1/3} D = 32$m (say 100 ft). This is the distance that applies if *one* gun is fired; for the broadside from a three-gun turret of figure 4.1 the blast distance is estimated to be about 50 m.

NOTE 15. EXTERNAL BALLISTICS 101: NO DRAG, NO LIFT

With no drag forces, it is easy to calculate the trajectory of a projectile. Consider figure N15, which shows the geometry: our projectile is launched at speed v_0 at angle θ_0. If the only force that acts upon it is gravity, then the horizontal and vertical components of speed at time t after launch is

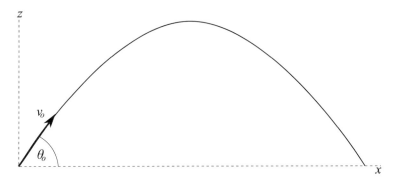

Figure N15. A projectile subjected only to the force of gravity has a parabolic trajectory if the range is short. The muzzle velocity of the projectile is defined by the initial elevation angle, θ_0, and the initial speed, v_0.

$$\dot{x} = v_0 \cos(\theta_0),$$
$$\dot{z} = v_0 \sin(\theta_0) - gt. \tag{N15.1}$$

The distance moved along the ground is

$$x(t) = v_0 t \cos(\theta_0), \tag{N15.2}$$

whereas the height above the ground is

$$z(t) = v_0 t \sin(\theta_0) - \frac{1}{2} gt^2. \tag{N15.3}$$

This height is zero at launch time $t = 0$ and at time τ, when the projectile lands back on the ground. (I am assuming that the ground is horizontal.) Thus,

$$\tau = \frac{2v_0}{g} \sin(\theta_0). \tag{N15.4}$$

The range is given by $R = x(\tau)$; from equations (N15.2) and (N15.4) we obtain

$$R = \frac{v_0^2}{g} \sin(2\theta_0). \tag{N15.5}$$

The shape of the trajectory becomes apparent when we eliminate t from equations (N15.2) and (N15.3):

$$z(x) = x \left(1 - \frac{x}{R} \right) \tan(\theta_0), \tag{N15.6}$$

which is a parabola.

NOTE 16. SIGHTS

There are two problems to solve for the case of rifle sight settings, as illustrated in figure 4.3. First, we want to know to what maximum range the bullet trajectory differs from the line of sight by less than d. For the flat trajectory of a rifle, we can assume that elevation angle θ is very small, so that from equations (N15.5) and (N15.6)

$$z(x) \approx \frac{g}{2v_0^2} x(R - x).$$ (N16.1)

I am using the same notation here as in technical note 15. From figure 4.3a you can see that the deviation is less than d for the whole trajectory so long as the following inequality is satisfied: $z(\tfrac{1}{2}R) \le 2d$. The maximum range corresponds to the equality. Substituting from equation (N16.1) and inverting, we obtain for the maximum range

$$R = 4v_0 \sqrt{\frac{d}{g}} \, .$$ (N16.2)

The second problem is to determine the range swath for which the sight setting is valid. That is to say, at ranges greater than that of equation (N16.2) the bullet may deviate from the line of sight by a distance exceeding d at some points of the trajectory, but it will be within d for a range swath of extent r. To determine r, note that the launch angle for small d is approximately given by

$$\theta_0 \approx \frac{d}{r} \, .$$ (N16.3)

Substituting in equation (N15.5) and inverting yields the desired formula:

$$r \approx \frac{2v_0^2}{g} \frac{d}{R} \, .$$ (N16.4)

Note that r increases as the square of muzzle speed.

For this calculation I have assumed that the target and the shooter are on the same level. If this is not so, then the results change. For example, if the target is uphill of the shooter, then the range swath r will be bigger than predicted by equation (N16.4); it will be smaller if the target is downhill. You can see why this is so by considering the trajectory shape in figure 4.3b.

NOTE 17. AERODYNAMIC DRAG

Here we investigate the influence of aerodynamic drag upon projectile trajectory, making use of the simplifying assumption that the drag coefficient is

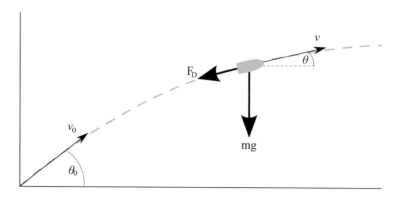

Figure N17. Aerodynamic drag force, F_D, acts in the opposite direction to projectile velocity, v. The projectile weight is mg.

constant. (This is the level 1 analysis of chapter 4.) From the geometry of figure N17 you can see that

$$m\ddot{x} = -F_D \cos(\theta),$$
$$m\ddot{z} = -F_D \sin(\theta) - mg. \tag{N17.1}$$

I think it is easier in this case to work with different variables: in terms of projectile speed $v = \sqrt{\dot{x}^2 + \dot{z}^2}$ and velocity direction θ we find that equations (N17.1) become

$$\dot{v} = -b_D v^2 - g \sin(\theta),$$
$$v\dot{\theta} = -g \cos(\theta). \tag{N17.2}$$

where

$$b_D = \frac{c_D \rho A}{2m}. \tag{N17.3}$$

To obtain equation (17.2) I have substituted for drag force from equation (4.1). The constant b_D (let's call it the "drag factor") has units of inverse length. Equations (N17.2) immediately give us the terminal speed of the projectile:[9]

$$v_T = \sqrt{\frac{g}{b_D}}. \tag{N17.4}$$

9. Terminal velocity is that projectile velocity for which the projectile has no net force acting on it. From equation (N17.2) this velocity is in the direction $\theta = -90°$ (straight down) with speed v_T, given in equation (N17.4).

For a skydiver this speed is about 53 ms⁻¹ (120 mph); for a NATO 7.62×5 bullet it is about twice as much.

For a short- or medium-range bullet trajectory we can assume that the velocity direction angle θ is small, in which case the equations (N17.2) can be solved to yield

$$v(t) \approx \frac{v_0}{1 + b_D v t}, \quad \theta(t) \approx \theta_0 - \frac{g}{v_0} t - 1/2\, b_D g t^2. \tag{N17.5}$$

By eliminating the variable t from equations (N17.5) we obtain $v\,(\theta)$; expressing this quantity in terms of horizontal distance x and height z (see fig. N17) gives us the trajectory curve:

$$z(x) \approx \frac{g}{(2b_D v_0)^2} \left[1 - \exp(2b_D x) + 2b_D x \left(1 + 2\, \frac{\theta_0 b_D v_0^2}{g} \right) \right]. \tag{N17.6}$$

This expression reduces to the equation obtained earlier (N15.3) in the limit of zero drag ($b_D \rightarrow 0$). Equation (N17.6) is plotted in figure 4.9.

The range R is the x value for which the projectile has returned to earth (assuming, as usual, that the ground is level); thus, R is defined via $z(R) = 0$. In general, it is difficult to obtain a closed-form expression for R from equation (N17.6), but we can do so for the case of short trajectories, for which $b_D x \ll 1$. With this restriction we calculate

$$R \approx \frac{2\theta_0 v_0^2}{g} \left[1 - \frac{1/3}{1 + \dfrac{g}{4\theta_0 b_D v_0^2}} \right]. \tag{N17.7}$$

Equation (N17.7) holds for short-range and flat trajectories only—say, for $R \leq$ 450 m and $\theta_0 \leq 0.2°$. For realistic parameter values, equation (N17.7) results in projectile ranges that are about two-thirds or three-quarters the value obtained for the same initial velocity in a vacuum.

The time of flight, τ, for a trajectory can be found by integrating the speed [eq. (N17.5)] and inverting. This gives us

$$\tau = \frac{1}{b_D v} [\exp(b_D R) - 1]. \tag{N17.8}$$

From equations (N17.5) and (N17.8) we can determine the bullet speed at the end of the trajectory:

$$v(\tau) = v_0 \exp(-b_D R). \tag{N17.9}$$

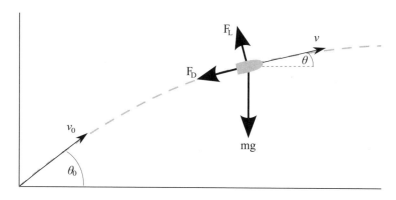

Figure N18. Aerodynamic lift force, F_L, acts perpendicularly to projectile velocity, v. Here F_L is directed upward, which is appropriate for arrows. For bullets, the lift force can act in a different direction.

NOTE 18. AERODYNAMIC LIFT

We now add a third force acting upon a projectile; figure N18 shows a projectile following a trajectory and acted upon by the force of gravity and by the forces of aerodynamic drag and lift. The analysis is similar to that of technical note 17, so I don't need to repeat it here. The resulting equations of motion are

$$\dot{v} = -b_D v^2 - g \sin(\theta)$$
$$v\dot{\theta} = b_L v^2 - g \cos(\theta) \tag{N18.1}$$

In the second of equation (N18.1) the lift factor, b_L, is defined in the same way as drag factor, shown in equation (N17.3), except that the lift coefficient, c_L, replaces the drag coefficient, c_D. The general solution to equations (N18.1) must be obtained by brute-force integration—computer number-crunching—as there is no analytic solution. As in technical note 17, however, we can readily obtain the projectile's terminal velocity from equation (N18.1). It has an elevation angle θ_T and a speed v_T given by

$$\sin(\theta_T) = \frac{-b_D}{\sqrt{b_D^2 + b_L^2}} \ , v_T = \left(\frac{g^2}{b_D^2 + b_L^2} \right)^{1/4}. \tag{N18.2}$$

So the terminal velocity is no longer directed vertically downward.

If we restrict our attention to flat trajectories with a small launch angle (appropriate for small-arm or ordnance trajectories but not for the trajectories of missiles such as arrows), then we can proceed as in technical note 17 to obtain an approximate expression for $v(t)$, equation (17.5), and the following expression for trajectory shape:

$$z(x) \approx \frac{g}{(2b_Dv_0)^2}\left[1 - \exp(2b_Dx) + 2b_Dx\left(1 + 2\,\frac{\theta_0 b_D v_0^2}{g}\right)\right] + 1/2\,b_Lx^2.$$

$$(N18.3)$$

This is the same as equation (N17.6) except for the last term.

NOTE 19. EARTH'S CURVATURE

Equation (N15.5) for projectile range applies only for short ranges. For longer ranges (tens of miles) we must include the effects of the earth's curvature. The trajectory is now that of figure N19; gravity acts toward the center of the earth (and so changes direction as the projectile moves). This problem is trickier to solve; to simplify the derivation, let's once again remove the earth's atmosphere. With no atmosphere there is no dissipative drag force, and so both angular momentum and energy are conserved. In the notation of figure N19,

$$mr^2\dot{\beta} = mv_0R_e\cos(\theta_0), \text{ where } \dot{\beta} = \frac{v}{r}\cos(\theta) \qquad (N19.1)$$

$$1/2m\dot{r}^2 + 1/2\,mr^2\dot{\beta}^2 - \frac{GMm}{r} = 1/2\,mv_0^2 - \frac{GMm}{R_e}. \qquad (N19.2)$$

The mean radius of the earth is $R_e = 6371$ km. The projectile location at any given instant is specified by its distance, r, from the center of the earth and the angle ω of figure N19. G is the universal gravitational constant, and M is the earth's mass; the product can be written as $GM = gR_e^2$. From the foregoing we can write down the radial equation of motion:

$$\dot{r} = \sqrt{v_0^2 - 2gR_e + \frac{2gR_e^2}{r} - \frac{v_0^2R_e^2\cos^2(\theta_0)}{r^2}}. \qquad (N19.3)$$

Note that $\dot{\beta} = \dot{r}d\beta/dr$ and substitute from equation (N19.3) to obtain

$$\frac{d\beta}{dr} = \frac{v_0R_e\cos(\theta_0)}{r\sqrt{(v_0^2 - 2gR_e)r^2 + 2gR_e^2r - v_0^2R_e^2\cos^2(\theta_0)}}. \qquad (N19.4)$$

Both equations (N19.3) and (N19.4) can be integrated analytically. (I won't write down the solutions because they are extremely tedious.) The solutions are substituted into the expression for projectile range, $R = 2R_e\beta(r_0)$, where r_0 is the value of r at maximum projectile height (see fig. N19). This value is readily found from equation (N19.3) by setting $\dot{r} = 0$. After tedious manipulations we arrive at the following equation for projectile range, R:

$$\sin\left(\frac{R}{2R_e}\right) = \frac{s\sin(\theta_0)\cos(\theta_0)}{\sqrt{1 - s(2 - s)\cos^2(\theta_0)}}, \text{ where } s \equiv \frac{v_0^2}{gR_e}. \qquad (N19.5)$$

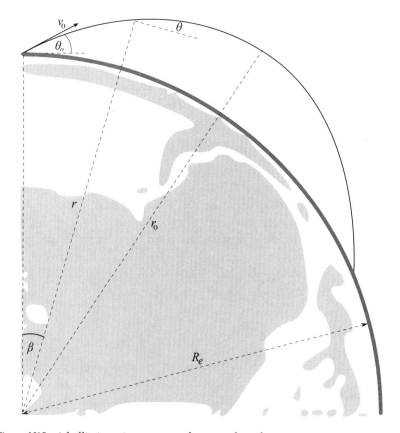

Figure N19. A ballistic trajectory over the curved earth.

For realistic projectile speeds we expect $s \ll 1$, and so we can expand equation (N19.5), neglecting terms of order s^2. The result is[10]

$$R \approx \frac{v_0^2}{g} \sin(2\theta_0) \left[1 + \frac{v_0^2}{gR_e} \cos^2(\theta_0) \right]. \tag{N19.6}$$

Note that, in the limit $R_e \to \infty$, we recover the flat-earth expression (N15.5). From equation (N19.6) it is easy to see that the maximum range no longer occurs for $\theta_0 = 45°$, but instead occurs for θ satisfying the following equation:

$$\cos(2\theta_0) = \frac{v_0^2}{2gR_e}. \tag{N19.7}$$

For example, if $v_0 = 2,000 \text{m}/\text{s}$, then the maximum range occurs for $\theta_0 = 44.1°$.

10. This result is also obtained by Thomson (1986, chap. 4) via a different calculation.

NOTE 20. ROCKETS

In this technical note we examine some basic rocket dynamics and derive the equations behind the graphs of figure 6.9. A standard undergraduate physics tutorial example derives the following equation for rockets from momentum conservation:

$$M \, dv = -u \, dM. \qquad (N20.1)$$

Here M is rocket mass (including fuel), dv is the change in rocket speed, and u is exhaust gas speed. We can generalize this equation as follows:

$$M \frac{dv}{dt} = F_{ext} - u \frac{dM}{dt}. \qquad (N20.2)$$

In this equation F_{ext} is any external force that acts upon the rocket, such as gravity or aerodynamic drag. The left side of equation (N20.2) is the rocket thrust. With no external force—say our rocket is in space—this equation can be integrated to yield rocket mass as a function of speed:

$$M(v) = M_0 \exp(-v/u). \qquad (N20.3)$$

Here, M_0 is the initial rocket mass. This equation is plotted in figure 6.9a. Note that the only parameter is gas ejection speed.

Now I will assume that the rocket motor ejects gas at a constant rate, $\dot{M}$, so that

$$M = M_0 - \dot{M}t \qquad (N20.4)$$

and also that our rocket is flying through the air, so that an external drag force acts. Thus, equation (N20.2) becomes

$$(M_0 - \dot{M}t) \frac{dv}{dt} = -b_D v^2 + u\dot{M}. \qquad (N20.5)$$

The drag factor b_D is assumed to be constant. Integrating we obtain (after some calculation) an expression showing how rocket speed increases with time:

$$\frac{v(t)}{u} = \frac{1 - (1 - \epsilon t/T)^{2u/c}}{1 + (1 - \epsilon t/T)^{2u/c}} \frac{c}{u}. \qquad (N20.6)$$

Here, T is the burn time for the rocket, $\epsilon = \dot{M}T/M_0$ is the fraction of rocket mass taken up by fuel, and c is the maximum possible rocket speed, given by

$$c = \sqrt{\frac{u\dot{M}}{b_D}}. \qquad (N20.7)$$

Equation (N20.6) is plotted in figure 6.9b for three values of the parameter c (0.5, 1.0, and 1.5). Note that now, in contrast to the case with no aerodynamic drag, rocket speed depends upon the rate, $\dot{M}$, at which gas is ejected as well as upon gas ejection speed, u. Eliminating the time, t, from equations (N20.4) and (N20.6) gives us rocket mass as a function of speed:

$$M(v) = M_0 \left(\frac{c - v}{c + v} \right)^{c/2u}.$$ (N20.8)

Equation (N20.8) is plotted in figure 6.9a for $c = 0.5$, $c = 1.0$, and $c = 1.5$.

Consider now a bullet that has the same "payload" mass as the rocket, that is, $(1 - \epsilon)M_0$, and the same drag characteristics. If we assume horizontal flight, its equation of motion is

$$(1 - \epsilon)M_0 \frac{dv_b}{dt} = -b_D v_b^2.$$ (N20.9)

I have attached a subscript b to the speed to indicate that this is bullet speed and not rocket speed. Equation (N20.9) integrates to

$$\frac{v_b(t)}{u} = \frac{c^2}{AC^2 + \dfrac{u^2 t}{AT}}, \text{ where } A = \frac{1 - \epsilon}{\epsilon}.$$ (N20.10)

Equation (N20.10) is plotted in figure 6.9b. From the intersections of $v(t)$ and $v_b(t)$ I have plotted figure 6.9c, which shows the fuel fraction ϵ required to make bullet speed equal to rocket speed at time T. For higher ϵ values, the rocket is faster.

NOTE 21. LRP PENETRATION DEPTH

Here is a simple argument providing a rough estimate of penetration depth for a very-high-speed projectile striking a target. The argument is based on momentum conservation and makes no reference to projectile energy or to the material properties of the target. We assume that because projectile speed is so high, target cohesion (the strength of the molecular bonds that bind the target material together) is unimportant and so the only target mass that acts to stop the projectile is that directly in front of it. The missile is stopped when it has pushed its own mass of target material out of the way (just as a pool ball stops when it hits another ball full on). The displaced target mass moves with the same speed as the projectile (the same as for the pool ball). So, target mass equals displaced projectile mass. But we are assuming that the cross-sectional area of displaced target mass is the same as the projectile cross-sectional area.

So, the length of displaced target, d, multiplied by target density, ρ_t, equals projectile length, l, multiplied by projectile density, ρ_p. The length of displaced target is just the penetration depth, which we have just estimated to be given by

$$d = \frac{\rho_p}{\rho_t} l. \tag{N21.1}$$

This argument is approximately valid so long as the target is not held together by high-tensile fibers and the missile nose is blunt. Given these assumptions, you can see that penetration depth increases if projectile length is increased or projectile density increases. This is why LRPs are long and made of dense material.

NOTE 22. EMPIRICAL ARMOR PENETRATION DEPTH FORMULAS

Lambert's formula describes the velocity needed to penetrate a steel plate:

$$v_{50} = 4000 \left(\frac{l}{c} \right)^{0.15} \sqrt{\frac{c^3}{m} \, [f + \exp(-f) - 1]} \, , \text{ where } f = \frac{d}{c \cos(\theta)}. \tag{N22.1}$$

In equation (N22.1), l, c, and d are, respectively, projectile length, projectile diameter (caliber), and penetration depth, all expressed in centimeters. The parameter m is projectile mass in grams. The angle at which the projectile strikes the plate is θ, where $\theta = 0$ corresponds to the projectile velocity being perpendicular to the plate—a head-on collision. Substituting for these parameters yields the required impact speed, v_{50}, in units of ms^{-1}. This is the speed required for 50% of projectiles to penetrate the plate to depth d. The graph of figure 7.5 is plotted from equation (N22.1).

If the projectile is an LRP missile, it can penetrate a depth of armor that is much greater than its diameter, so that $f \gg 1$. In this case equation (N22.1) can be simplified and written in the form

$$d \approx \left(\frac{v_{50}}{1035} \right)^2 c^{0.30} l^{0.70}. \tag{N22.2}$$

Again, speed is expressed in ms^{-1} and lengths in centimeters. Equation (N22.2) is plotted in figure 7.6 for two LRP projectiles.

There are many empirically derived formulas for armor penetration depths that apply to different projectile shapes or ranges of speeds (which may overlap).[11] The venerable DeMarre formula gives the projectile kinetic energy required to penetrate a steel target:

$$E = kc^{0.75}d^{0.70}. \tag{N22.3}$$

This equation looks nothing like Lambert's formula, but it really is quite similar. For example, if we recall that kinetic energy is $E = \frac{1}{2}mv^2$, and if we express mass as the product of projectile density and volume (where volume is area multiplied by length), then substitution in equation (N22.3) gives the same relationship between penetration depth and projectile length as we saw in equation (N22.2).

11. For a comprehensive study of penetration dynamics, see Ben-Dor, Dubinsky, and Elperin (2005) and the voluminous references therein. Lambert's formula is discussed in, e.g., Grace (1999); Carlucci and Jacobson (2008, chap. 15) discuss DeMarre's formula.

Glossary

Ballistics is a widespread and ancient science and consequently has evolved, over the years, a language of its own. A comprehensive glossary would therefore be as long as a dictionary. Here I include only ballistics terms that are used in the book, and even then I will be selective, omitting words of peripheral or of obvious and unambiguous meaning. Thus, for example, you will find *drag force* but not *gravitational force*, and *smart bomb* but not *bomb*.

Angle of attack. The angle between an airfoil (such as a bullet) and the wind direction.

Archer's paradox. An apparent paradox concerning the flight of a longbow arrow as it passes the bow handle. Geometry suggests that the arrow flight should be deflected, but physics ensures that it is not.

Arquebus. An early shouldered firearm, a muzzle-loading smoothbore with a wheel-lock firing system; ancestor of the musket.

Artillery. Cannon; the ordnance branch of an army.

Assault rifle. A military rifle capable of both automatic and semiautomatic fire, of intermediate power, usually with a large magazine.

Assegai. A throwing spear associated with the Zulus.

Atmospheric attenuation. The exponential reduction in the density of air with height above the earth's surface.

Automatic. A firearm that reloads and fires rapidly and repeatedly until the trigger is released.

Ball. Spherical musket ammunition. The name was retained even for nonspherical ammunition such as the Minié bullet and provides us with the term "round."

Ballistic coefficient. The sectional density of a bullet divided by its form factor—a complicated and outdated, but popular, measure of bullet "slipperiness," or ability to pierce the air.

Ballistic gelatin. A substance used as a substitute for human flesh in investigations of the effects of bullet penetration.

Ballistic pendulum. An eighteenth-century invention for measuring the speed of a bullet, consisting of a block of known mass (into which the bullet is fired) suspended from two rods that are free to rotate. Bullet speed can be calculated from the measured block deflection.

Ballistics. The study of the dynamics of unguided (i.e., free-flying) projectiles.

Barrel. The tube through which a bullet travels when fired from a gun. The barrel determines bullet direction and spin.

Base bleed. Gas jet emission from the base of an artillery shell, which reduces tail drag by filling in the partial vacuum that forms behind the base.

Battery. A tactical unit of several artillery weapons and their crews, used for combined action.

Beamrider. A guided missile that flies along the direction of a guidance beam that is pointed at the missile target.

Benchrest. A tablelike support for a target rifle. Benchrest shooters are motivated chiefly by maximizing accuracy at long range, and thus customize their rifle and ammunition to attain this end.

Berdan. A cartridge primer system popular outside the United States. Slightly less expensive to produce than the rival Boxer system but not reusable.

Black powder. A low explosive used for centuries as the propulsive charge for firearms, consisting of a fine powder that is a mixture of charcoal, sulfur, and potassium nitrate. Superseded by smokeless powder.

Blowback. An automatic pistol mechanism powered by recoil and gas pressure.

Boat tail. A tapered bullet shape that reduces tail drag.

Bodkin. An armor-piercing arrowhead, bullet-shaped rather than flat.

Bolt. A crossbow arrow, considerably shorter than a longbow arrow.

Bombard. An ancient type of cannon with a short barrel and large caliber that fired stone or iron shot in a high-angle trajectory. Ancient equivalent of the mortar.

Bombardier. The crew member on a bomber airplane responsible for bomb aiming and release. Originally, any member of a gun crew.

Bore. The hollow part of a gun barrel.

Bourrelet. The largest-diameter parts of an artillery shell, excluding the driving band.

Boxer. The cartridge primer system most used within the United States, favored because it can readily be reused. See also *Berdan*.

Breech. The rear part of the barrel into which the cartridge is inserted.

Breech loader. A gun or small arm which is loaded via the breech.

Brisance. A measure of the rapidity of an explosion or deflagration.

Bullpup. A modern assault rifle design in which the action and the magazine are behind the trigger, so that part of the barrel is within the stock. This design shortens the weapon without shortening the barrel.

Caisson. A two-wheeled wagon containing ammunition boxes, attached to the limber of a horse-drawn artillery piece.

Caliber. The inside diameter of a smoothbore barrel; for a rifled barrel, either the distance between opposite grooves or opposite lands.

Cannon. A (usually) large-caliber gunpowder weapon. An artillery piece.

Caplock. A mechanism that used a percussion cap to ignite propellant charge. Succeeded the flintlock.

Carbine. A light, short-barreled rifle (often adopted by cavalry).

Carronade. A smoothbore cannon in the Age of Sail, short and usually of large caliber. Because carronades were light (so requiring a small charge, resulting in a short range), they could be deployed on the upper decks of ships.

Cartridge. A brass or copper case containing primer, propellant charge, and bullet. Historically, cartridges were rolls of paper containing powder and shot. Cartridges are also known as "rounds," harking back to the days of musket balls.

Cavitation. The creation of an opening in a wound due to the passage of a bullet. Temporary cavities can be very large, but close over quickly. Permanent cavities—the bullet trail resulting from tissue damage—are more closely correlated with wound severity.

Centerfire. A cartridge with the primer in the center of the base.

Center of gravity (CG). The point within an extended object at which all the mass can be considered to concentrate. Sometimes called "center of mass."

Center of pressure (CP). The point within an extended object at which aerodynamic forces can be considered to act.

Centrifugal force. A fictitious force that acts in a direction pointing away from the center of rotation.

Charcoal. The element carbon in a porous form derived from wood, which forms 15% of the black powder mixture.

Charge. The propellant powder within a cartridge or shell (or poured down the barrel of a musket).

Chobham armor. Layered, composite MBT armor consisting of small ceramic bricks and steel. More resistant to HEAT rounds than monolithic steel armor.

Chronograph. An instrument for measuring the time it takes a projectile to pass from one point to another, from which the projectile speed can be determined. Successor to the ballistic pendulum.

Closed-loop system. In control theory, a system with feedback. A closed-loop guidance system receives information back from the environment that informs the system of any guidance errors, which can then be corrected.

Coanda effect. A fluid dynamical effect in which a jet of air (or any fluid) tends to follow the curve of a smooth surface. Water from a faucet, poured over the back of a spoon, will follow the curve of the spoon instead of flowing vertically—try it and see.

Cock. The hammer of a flintlock weapon. Also the action of setting the firing mechanism into position for firing. Revolvers often have an intermediate half-cock position, as did flintlock weapons.

Composite bow. A bow made of different materials in layers, glued together. Typically, a wooden core is strengthened by horn on the belly (the inside curve of the bow, nearest the archer) because horn is strong in compression, and by sinew on the back (the outside curve) because sinew is strong in tension. A composite bow is smaller than a self bow of the same draw weight.

Computational fluid dynamics. A branch of physics that uses numerical methods (i.e., computer number-crunching) to solve the intractable equations of fluid dynamics, a field that includes ballistic aerodynamics.

Coriolis force. A fictitious force that depends upon projectile velocity. It acts in a direction that is perpendicular to both the velocity vector and the axis of rotation (of the earth, for ballistic projectiles).

Corned powder. Granular black powder. Grain size was varied to control deflagration rate. Corning probably originated to reduce moisture absorption, which rendered the propellant useless.

Counterpoise siege engine. A siege engine, such as the trebuchet, that was powered by gravity. Counterpoise engines were a product of the Middle Ages.

Crossbow. A short, powerful bow mounted on a stock. Often spanned (drawn) mechanically.

Culverin. Originally a portable firearm, held under the arm, the culverin was little more than a hollow iron tube. Later, culverins developed into heavier swivel guns used on board ships.

Dead ground. Terrain on a battlefield that cannot be seen or reached by artillery or small arms fire.

Deflagration. Burning at a rate that is slower than an explosive's detonation.

Degree of freedom (DoF). In ballistics, one of the different ways that a projectile can move. There are six DoF: translation (linear movement) in three directions (up and down, sideways, or forward and backward) and rotation about each of these directions.

Degressive burn. A propellant burn that slows with time, as the surface area of unburned propellant gets smaller (e.g., for a cylindrical rod, or a sphere).

Detonation. Combustion of an explosive fast enough to create a supersonic shock wave. Detonation is too fast for a propellant; explosive propellants would burst a gun barrel.

Dive bomber. A military plane, important in World War II, that dives steeply upon its surface target and thus guides its bomb before release. Dive bombers were much more accurate than conventional bombers.

Double base. A rapidly burning powder such as Cordite or Ballistite, made from nitroglycerin and guncotton.

Drag coefficient. A dimensionless quantity that defines the magnitude of the aerodynamic drag force acting upon a projectile. Drag coefficient depends upon a projectile's shape and texture.

Drag force. The component of air resistance that opposes the projectile velocity.

Drift. The lateral movement of a spinning projectile. In the absence of drift, a projectile's trajectory is confined to a vertical plane that includes the line of sight. Drift may also be due to a cross wind.

Driving band. The raised band of an artillery shell, made of soft metal, that engages the rifling and acts as a seal. Also known as "rotating band."

Dynamic stability. The property of a bullet trajectory in which the bullet's nose tends to align with the velocity direction and changes as the velocity direction changes.

External ballistics. The ballistics of a projectile after it has cleared the barrel; the study of ballistic trajectories through the atmosphere.

Fictitious force. No, not a figment of the imagination. A fictitious force is one that arises in a rotating system as felt by an observer within the system. Centrifugal and Coriolis forces are fictitious. An observer outside the system does not feel the fictitious forces that arise from the rotation.

Fineness ratio. The ratio of projectile (or projectile nose cone) length to radius. Describes the overall shape of a projectile.

Fire-and-forget missile. A missile that is capable of guiding itself to its designated target.

Firearm. Any weapon (sometimes only a small arm) that uses gunpowder to propel a projectile.

Fire control. In ballistics, controlling the aim of a gun or gun battery so that gunfire will strike a designated target. In particular, aiming off the line of sight, to counter the effects of gravity, drift, and other forces.

Flechette. A small steel dart with fins used as antipersonnel projectiles and distributed by the thousands at high speed from a bursting artillery shell. Modern shrapnel.

Flight arrow. An arrow designed to achieve maximum range by maximizing lift and minimizing drag.

Flintlock. A mechanism for igniting primer charge by causing a piece of flint to strike a frizzen and generate sparks. A firearm with a flintlock mechanism.

Form factor. The drag coefficient of a bullet divided by the drag coefficient of a standard reference bullet.

Fragmentation. The splitting up of an antipersonnel round by explosion (in an artillery shell) or contact (in a bullet), leading to the creation of high-speed fragments and an increased number of casualties or greater wound damage.

Frizzen. A piece of steel on the lid of a flintlock priming pan which, when struck by the flint, generated sparks and ignited the primer.

Full metal jacket. A bullet (usually of lead) encased in a shell of harder metal. The shell prevents lead from depositing on the rifle barrel. Such deposits can foul the barrel; they also cause marks on the bullet that can adversely influence accuracy.

Fuze. An electrical or mechanical mechanism in an artillery shell (or a bomb or a mine) that detonates the charge at a preselected time or when in the immediate proximity of a target.

Gatling gun. An early machine gun with a revolving cluster of barrels that are fired in sequence as the cluster is rotated.

Grain. A measure of propellant weight and of bullet weight equaling $\frac{1}{7,000}$ pound. Also a small particle of powder.

Guided missile. A missile that is guided to its target by responding to external information. This information may take many forms. For example, it may consist of instructions sent along a wire by the missile launcher or target location coordinates obtained from the missile's own remote sensor (such as heat detected in an airplane's jet plume).

Gun. A long-barreled cannon with a low angle of fire. Informally, any firearm.

Gunpowder. Today, any of the smokeless powder propellants. Formerly the name for black powder.

Gyroscopic effect. The tendency for a fast-spinning mass to resist forces that act to change the orientation of the spin. The change in direction that does occur is perpendicular to the direction of the applied force, often leading to precession and nutation.

Half cock. An intermediate position of a firearm hammer. In a flintlock weapon, the hammer was pulled back partway to permit priming the pan, but not the whole way, to prevent accidental discharge. Later weapons retained the half-cock position for safety reasons.

Handload. To assemble firearm cartridges from components, instead of purchasing fully assembled cartridges. The noun form has come to mean the cartridges assembled by handloading.

Heave. Linear motion up and down, one of the six degrees of freedom of a ballistic projectile.

High explosive (HE). A powerful explosive, one with high brisance.

High-explosive antitank (HEAT). An antitank projectile with a shaped-charge warhead.

Howitzer. A large-caliber cannon intermediate between gun and mortar.

Hypervelocity. An overblown way of saying "very fast."

Ignition. The process of igniting a propellant; the initial combustion.

Induced drag. The component of aerodynamic drag associated with aerodynamic lift. If an airfoil or projectile adopts a position with respect to the airflow that generates lift, the corresponding increase in aerodynamic drag is termed "induced drag."

Infrared (IR). A portion of the electromagnetic spectrum, contiguous with the visible spectrum at longer wavelengths. IR radiation is felt as heat.

Intercontinental ballistic missile (ICBM). A missile that follows a mainly ballistic trajectory, much of which is in the attenuated upper atmosphere.

Internal ballistics. The dynamics of a bullet or shell while it is still inside the gun barrel.

Javelin. A light throwing spear.

Lambert's formula. One of several empirical formulas for estimating the median projectile speed required to penetrate a given depth of steel. The formula is expressed in terms of projectile dimensions and mass, and impact angle.

Lands. The uncut part of a barrel after the rifling grooves have been cut. The lands are in contact with the bullet as it is propelled down the barrel.

Lift force. The component of air resistance that is directed perpendicularly to projectile velocity.

Limber. A two-wheeled wagon used for hauling cannon in the days of horse-drawn artillery.

Longbow. A long hand-drawn bow.

Long rod penetrator (LRP). A modern arrow; a very fast rod made of dense metal that penetrates tank armor.

Machine gun. A fully automatic firearm that fires rifle bullets in quick succession from a large magazine.

Mach number. The speed of a projectile through the air divided by the speed of sound through the same air.

Magazine. A container that holds ammunition in position so that it can be readily loaded, one round at a time, into the chamber.

Magnus force. An aerodynamic force that acts upon a spinning object. The Magnus force is directed perpendicularly to both the projectile spin axis and its velocity direction.

Main battle tank (MBT). A heavy tank.

Matchlock. The earliest mechanism for firing a hand-held firearm. A slow-burning string was presented to the primer via a lever mechanism connected to the trigger.

Median. In statistics, the value of a distribution for which half the samples have a lower value and half have a higher value.

Meplat. The nose of a bullet. "Meplat" refers to the bullet's shape, not its structure, and is an important aerodynamic consideration.

Minié ball. The first successful cylindro-conical bullet. The design made breech-loaded weapons practical by reducing windage and loading time.

Mitrailleuse. A multibarreled rifle that fired rounds in rapid succession from different barrels. A precursor of the true machine gun.

Mortar. An indirect-fire (i.e. high-angle trajectory), short-barreled, muzzle-loaded weapon.

Multiple round simultaneous impact (MRSI). In ballistics, an artillery technique that takes advantage of the different flight times of trajectories with different elevation angles of fire. The trajectories are sequenced so that several long-range rounds arrive at the target simultaneously. This may be tactically advantageous, in catching the enemy unprepared.

Munroe effect. The focusing of blast energy to take advantage of the geometry of a shaped charge. Also known as the "hollow-charge effect." It is the Munroe effect that makes RPG missiles so effective against armored vehicles.

Mushrooming. In ballistics, the flattening of a soft-nosed bullet upon impact with flesh, causing greater wound damage than would be the case for a bullet of the same speed and caliber that did not mushroom.

Musket. A type of shoulder gun: a long-barreled, muzzle-loaded smoothbore. Generally used to refer to the most advanced of these weapons.

Muzzle. The front of a gun barrel.

Muzzle brake. A slotted device that deflects escaping propellant gas, thus reducing the recoil or kick of a firearm. Particularly important for artillery.

Muzzle loader. A firearm that is loaded via the muzzle.

Muzzle velocity. The speed (not velocity) at which a projectile leaves the muzzle.

Neutral burn. A propellant burn that is constant and does not vary with time. Flat flakes of propellant burn in this way.

Nock. The slot at the feathered end of an arrow that holds it in place on the bowstring. Also applies to the grooves on the ends of a bow that hold the bowstring in place.

Nose cone. The front of an airborne projectile or rocket. Nose cone design has importance for aerodynamic drag.

Nutation. The wobbling of the axis direction of a precessing object, so that the precession is no longer smooth, but is sinusoidal or spiraling.

Ogive. The tapered end of a missile; the shape of a nose cone, usually symmetric about its longitudinal axis.

Onager. A large catapult of classical antiquity, with a single arm that swings in a vertical plane, powered by twisted sinew.

Open-loop system. In control theory, a system without feedback. An open-loop guidance system proceeds according to a set of internal instructions that cannot be corrected or updated after launch.

Ordnance. Mounted guns, usually of larger caliber than small arms.

Overstabilized. Descriptive of a bullet or shell with too much spin, so that the spin axis does not change direction throughout the trajectory. This matters for high-angle trajectories because the gun elevation angle (which is also the spin direction angle) differs a lot from the projectile velocity direction near the end of the trajectory; thus, an overstabilized shell may land on its side.

Overturn. With respect to a nonspherical projectile, such as a bullet, to turn end over end because of lack of gyroscopic stabilization. A projectile that is not gyroscopically stabilized will be subject to an asymmetric force due to wind pressure acting at the projectile nose. This asymmetric force applies a torque that causes the projectile to overturn.

Parabola. The trajectory of a projectile on a flat earth with no air resistance. It is symmetric about its highest point.

Percussion cap. A cylinder containing mercury fulminate (or some other impact-sensitive explosive) that was used to set off propellant charge. Percussion caps replaced flints and black powder primer.

Pilum. A Roman throwing spear made of soft metal that embedded itself in an enemy infantryman's shield.

Pinfire cartridge. An obsolete cartridge with an external pin connected to the primer.

Pistol. A handgun; a firearm designed to be operated by one hand. Some people do not consider revolvers to be pistols; they define pistols as handguns in which the chamber is part of the barrel.

Pitch. Rotation about a lateral axis, causing an object to rock up and down. One of the six degrees of freedom.

Precession. The directional change of the axis of a spinning object. For projectiles, precession is a (very roughly circular) rotation about the velocity vector.

Pressure drag. The component of aerodynamic drag that is due to pressure differences over the surface of a projectile. Also known as "form drag."

Primer. A small, usually pressure-sensitive charge which is set off by the trigger, and which in turn ignites the main propulsive charge of a cartridge.

Profile drag. The sum of pressure drag and tail drag.

Progressive burn. A propellant burn that becomes faster with time because the propellant shape (a perforated cylinder) causes the area of burning surface to increase as it burns.

Projectile. Any object that moves under external forces only after launch. A sling stone, a bullet, and an artillery shell are projectiles. A rocket is not: it is a missile.

Reactive armor. Armor that reacts (usually by causing an explosion) to being struck by a projectile or missile, in such a manner as to reduce the damage that would otherwise be caused by the projectile or missile.

Recoil. The reaction force of a bullet or shell against the gun that fired it, due to the rapid expansion of propellant gas. Also known as "kick."

Recurve bow. A prestressed bow with ends that, when not strung, curve away from the archer. Recurved bows are more powerful than bows of the same size and material that are not recurved.

Reference frame. A set of axes, or coordinate system, in which properties of a system are measured. For example, we may set up a frame of reference in which the x-axis points east, the y-axis points north, and the z-axis points vertically up. Observers in two different reference frames that are accelerating relative to one another will not experience the same forces.

Revolver. A handheld firearm with a cylindrical magazine that rotates after firing to align the next chamber with the barrel.

Reynolds number. A dimensionless number that characterizes airflow conditions. Specifically, Reynolds number indicates the relative importance of inertial and viscous forces.

Rifle. A shoulder firearm that has a single rifled barrel capable of firing bullets one at a time.

Rifled musket. A muzzle-loading shoulder weapon with a rifled barrel. These weapons—more difficult to load than muskets but much more effective— were a transitional technology between muskets and rifles.

Rifling. Spiral grooves cut into the inside surface of a barrel to cause the bullet to spin.

Right-hand rule. A convention for determining the direction to be associated with the spin of an object: If the fingers of the right hand are curled in the path of rotation, then the direction of object spin is defined to be that of the extended thumb.

Rimfire cartridge. A cartridge with the pressure-sensitive primer placed in a rim at the base, now usually confined to .22 caliber rifle ammunition.

Rocket. A missile that carries its own propellant and obtains thrust by expelling propellant gas at high speed.

Rocket-propelled grenade (RPG). An inexpensive anti-armor weapon, firing self-propelled missiles that are often armed with a shaped charge. Bazookas and Panzerfausts were early RPG launchers.

Roll. Rotation about the direction of motion. One of the six degrees of freedom.

Sabot. A disposable carrier that positions a projectile within the barrel of a larger-caliber gun and acts as a seal for the propellant gas.

Saltpeter. Potassium nitrate, a naturally occurring chemical compound that forms 75% of the black powder mixture.

Sectional density. The mass (or sometimes weight) of a bullet divided by the square of its caliber.

Self bow. A bow made from a single piece of wood.

Semiautomatic. A firearm that fires one round per trigger pull but automatically reloads. The trigger action causes the next round to be positioned in readiness for firing.

Serpentine. Fine black powder, an early type of powder. The ingredients had to be ground finely to make the powder burn effectively, but it was soon learned that the burn rate could be controlled by corning the mixture. Also, an S-shaped element of a matchlock mechanism.

Shaped charge. An explosive charge shaped to direct its blast in a thin jet (see *Munroe effect*) so as to penetrate armor.

Shell. A usually large-caliber artillery round, carrying a payload of explosives with a fuze.

Shock wave. In ballistics, the pressure wave that accompanies a supersonic missile, typically exhibiting a very rapid change of pressure across the wave. Shock waves carry away energy and so contribute a component (increasingly important as missile speed increases) to aerodynamic drag.

Shot. Usually, a solid projectile fired from a cannon.

Siege engine. Any of several types of artillery before the age of gunpowder, which fired stone balls (or occasionally dead animals). Wooden catapults were powered by twisted sinew or by gravity. Some of the later examples grew very large.

Single base. A smokeless powder such as guncotton that is made from nitrocellulose only (i.e., with no explosive such as nitroglycerine added).

Skin drag. The component of aerodynamic drag that is caused by friction between the airflow and the surface of a projectile.

Sling. A projectile weapon consisting of two lengths of cord attached to a pouch, which holds the sling stone. Used properly, the sling acts as an extension of the arm, enabling the slinger to project the stone much further than he could throw it.

Small arm. Any of various Infantry weapons, capable of being carried and operated by one person.

Smart bomb. A guided bomb with control surfaces (fins), allowing it to change direction so that it can strike a target that has been "painted" (for example, by a laser beam directed from the bomber).

Smokeless powder. A nitrocellulose-based propellant that releases much less smoke than the black powder it replaced.

Smoothbore. A barrel with no rifling. Shotguns and some artillery weapons are smoothbore; until the last third of the nineteenth century, most firearms were smoothbore.

Sound and flash. A method of improving the accuracy of an artillery battery aiming at an enemy gun by utilizing the difference in time between the enemy gun flash and gun blast. Essentially, a triangulation technique.

Spaced armor. Secondary tank armor designed to cause a shaped charge to detonate prematurely, so that it dissipates its power before contacting the main armor of the tank. Also known as "slat armor."

Spalling. In ballistics, the flaking of material on the inside of an armored vehicle, directly adjacent to the external point of a missile strike. Spall (armor fragments) can cause considerable death and destruction within the armored vehicle.

Spin stabilization. The stabilization a bullet achieves by spinning, even when it is aerodynamically unstable. A spinning bullet will resist changing orientation (e.g., by overturning) because of the gyroscopic effect.

Spitzer. A sharp-pointed bullet. This shape can reduce drag and increase target penetration.

Static stability. The property of a bullet trajectory in which the bullet does not overturn or tumble—that is, its longitudinal axis changes direction only slowly.

Staff sling. A sling weapon that is attached to the end of a staff, thus increasing the lever arm advantage of the sling.

Standard deviation. A statistical measure of variation of a distribution. In ballistics, shots are distributed randomly about the intended target, especially at long range; the standard deviation is a measure of how widespread the shots are.

Subsonic. Moving at a speed that is below the speed of sound. See *Supersonic*.

Sulfur. A naturally occurring chemical element that forms 10% of the black powder mixture.

Supersonic. Moving at a speed that exceeds the speed of sound. In ballistics, the nature of air resistance for a projectile traveling above the speed of sound is very different from that of a projectile moving at a speed that is below the speed of sound.

Surge. Linear motion forward and backward. One of the six degrees of freedom.

Sway. Linear motion from side to side. One of the six degrees of freedom.

Tachometric bombsight. A bombsight (such as the Norden bombsight of World War II) that corrects bomb aiming for the measured wind speed between the bomber and its ground target.

Tail drag. The component of aerodynamic drag of a bullet or shell resulting from the shape of its base. Tail drag results from a partial vacuum that forms immediately behind the base.

Tandem charge. An anti-armor round designed to combat reactive armor. An initial charge causes the detonation of the target's reactive armor, and then the main (shaped) charge is able to penetrate the target's main armor.

Terminal ballistics. The ballistics of the final phase of a projectile, as it penetrates or attempts to penetrate its target.

Torque. A tendency to twist.

Tractable. In ballistics, descriptive of a bullet trajectory in which the bullet's longitudinal axis is aligned with its velocity vector throughout the trajectory. A tractable trajectory is desirable because it minimizes drag and so increases bullet range.

Trajectory. The shape of a projectile's flight path.

Transitional ballistics. The ballistics of a firearm projectile that has just left the gun barrel and so is subject to the blast force of escaping propellant gas.

Transonic. Moving at speeds that are close (within about 20%) to the speed of sound. The behavior of ballistic projectiles moving at transonic speeds is very difficult to analyze.

Trebuchet. A counterpoise siege engine that relied upon the force of gravity to propel its (sometimes very heavy) projectiles. The largest of the catapults, trebuchets were a product of the medieval world; they were unknown in classical antiquity.

Trigger. The firearm mechanism that actuates the firing pin.

Trunnion. Protrusions from the side of a barrel of an ancient cannon, permitting it to be aimed in elevation.

Tumbling. The overturning, end-over-end motion of an unstabilized bullet.

Turbulence. Chaotic fluid flow. In ballistics, characterizing airflow in the wake of a projectile that is largely responsible for tail drag.

Twist. A measure of the tightness of the rifling spiral: the distance along the barrel required for one complete turn of the spiral.

Unmanned air vehicle (UAV). A remotely controlled or autonomous aircraft used for surveillance and strike missions.

Vector. A mathematical representation of physical properties, such as velocity and force, that have both magnitude and direction.

Von Kármán ogive. A missile nose cone shape with the minimum drag for a given cross-sectional area.

Wave drag. The component of aerodynamic drag that is due to the presence of shock waves.

Wheel-lock. A firearm firing mechanism that succeeded matchlocks and preceded flintlocks. Wheel-locks ignited the primer by sparks from a rotating steel wheel.

Windage. The difference between bore diameter and bullet diameter. Windage was necessary for muzzle-loaded smoothbore firearms.

Wind force. In this book, the name given to the force of air resistance. It has two components: aerodynamic drag and aerodynamic lift.

Wire-guided missile. A missile that is guided by instructions delivered via a wire that is spooled out behind the missile, connecting it to the launcher.

Yaw. Rotation about a vertical axis, causing a projectile to rotate horizontally with respect to its direction of motion. One of the six degrees of freedom.

Yaw of repose. The remnant yaw angle of a spinning projectile after gyroscopic precession has been damped down.

Bibliography

Baden, G. P. 1961. *Development of Airborne Armament, 1910–1961.* Historical Division, Office of Information, Air Force Systems Command. AFSC Historical Publication Series 61-52-1, October. Available at www.alternate wars.com/SAC/AirborneArmament/Volume_I.htm.

Beevor, A. 2002. *Berlin: The Downfall, 1945.* London: Viking.

Ben-Dor, G., A. Dubinsky, and T. Elperin. 2005. "Ballistic impact: Recent advances in analytical modeling of plate penetration dynamics; A review." *ASME Applied Mechanics Reviews* 58:355–71.

Benton, Capt. J. G. 1862. *A Course of Instruction in Ordnance and Gunnery, Compiled for the Use of the Cadets of the United States Military Academy.* New York: van Nostrand.

Berger, M. L. 1979. *Firearms in American History.* New York: Franklin Watts.

Blackwood, J. K., and F. P. Bowden. 1952. "The initiation, burning, and thermal decomposition of gunpowder." *Proceedings of the Royal Society A* 213: 285–306.

Boos-Bavnbek, B., and J. Høyrup. 2003. *Mathematics and War.* Basel: Birkhäuser.

Bradbury, J. 1992. *The Medieval Archer.* Rochester, NY: Boydell and Brewer.

Britannica. 1998. *Encyclopaedia Britannica.* CD 98 Standard Edition.

Buchanan, B. J., ed. 2006. *Gunpowder, Explosives, and the State.* Aldershot, UK: Ashgate.

Canada DND (Department of National Defence). 1992. *Field Artillery.* Vol. 6, *Ballistics and Ammunition.* B-GL-306-006/FP-001. Available at http://armyapp .forces.gc.ca/ael/pubs/B-GL-371-006-FP-001.PDF.

Carlucci, D. E., and S. F. Jacobson. 2008. *Ballistics: Theory and Design of Guns and Ammunition.* Boca Raton, FL: CRC Press.

Chevedden, P. E., et al. 1995. "The trebuchet." *Scientific American* 273:66–71.

Chowdhary, A. G., and J. H. Challis. 2001. "The biomechanics of an overarm throwing task: A simulation model examination of optimal timing of muscle activations." *Journal of Theoretical Biology* 211:39–53.

Courtney, M., and A. Courtney. 2007. "The truth about ballistic coefficients." May 2. Available at http://arxiv.org/abs/0705.0389.

Cross, R. 2003. "Physics of overarm throwing." *American Journal of Physics* 72:305–12.

Denny, M. 2003. "Bow and catapult internal dynamics." *European Journal of Physics* 24:367–78.

——. 2005. "Siege engine dynamics." *European Journal of Physics* 26:561–77.

——. 2007. *Ingenium: Five Machines That Changed the World*. Baltimore: Johns Hopkins University Press.

——. 2009. *Float Your Boat! The Evolution and Science of Sailing*. Baltimore: Johns Hopkins University Press.

——. 2010. Mark Denny Web site. http://markdenny.shawwebspace.ca.
My papers about bow and catapult internal ballistics, onager internal ballistics, and trebuchet dynamics are available on this Web site.

DeVoto, J. G. 1993. *Flavius Arrianus: Technè Taktika (Tactical handbook) and Ektaxis kata Alanoon (The expedition against the Alans)*. Chicago: Ares.

Di Maio, V. J. M. 1999. *Gunshot Wounds: Practical Aspects of Firearms, Ballistics, and Forensic Techniques*. Boca Raton, FL: CRC Press.

Dougherty, P. J., and H. C. Eidt. 2009. "Wound ballistics: Minié ball vs. full metal jacketed bullet; A comparison of Civil War and Spanish-American War firearms." *Military Medicine* 174:403–7.

Farrar, C. L., D. W. Leeming, and G. M. Moss. 1999. *Military Ballistics: A Basic Manual*. London: Brassey's.

Farwell, B. 2001. *The Encyclopedia of Nineteenth-Century Land Warfare*. New York: Norton.

FITA (International Archery Association). 2009. "Flight records—men." September 10. www.archery.org/content.asp?id=892&me_id=753&cnt_id =895.
The International Archery Federation Web site provides details of current records for different classes of archery competition, including flight arrow ranges.

Fowler, K. 1967. *The Age of Plantagenet and Valois*. New York: Putnam.

Grace, F. I. 1999. "Ballistic limit velocity for long rods from ordinance velocity through hypervelocity impact." *International Journal of Impact Engineering* 23:295–306.

Green, M., and G. Stewart. 2005. *M1 Abrams at War*. Osceola WI: Zenith Press.

Greener, W. W. 2002. *The Gun and Its Development*. Guilford, CT: Globe Pequot. (Orig. pub. 1881.)

Guthrie, W. P. 2002. *Battles of the Thirty Years' War*. Westport, CT: Greenwood.

Hacker, B. C. 2006. *American Military Technology: The Life Story of a Technology*. Baltimore: Johns Hopkins University Press.

Hall, B. S. 1997. *Weapons and Warfare in Renaissance Europe*. Baltimore: Johns Hopkins University Press.

Hardy, R. 1993. *The Longbow*. New York: Lyons and Burford.

Hatcher, J. S. 1962. *Hatcher's Notebook*. Harrisburg, PA: Stackpole Books.

Heath, B. 2005. *Discovering the Great South Land*. Kenthurst, NSW, Australia: Rosenberg.

Heidler, D. S., and J. T. Heidler, eds. 2002. *Encyclopedia of the American Civil War: A Political, Social, and Military History*. London: Norton.

Heramb, R. M., and B. R. McCord. 2002. "The manufacture of smokeless powders and their forensic analysis: A brief review." *Forensic Science Communications* 4, no. 2.

Hess, E. J. 2008. *The Rifle Musket in Civil War Combat: Reality and Myth*. Lawrence: University Press of Kansas.

Hogg, I. V. 1970. *The Guns, 1939–45*. New York: Ballantine.

Kelly, J. 2005. *Gunpowder: Alchemy, Bombards, and Pyrotechnics; The History of the Explosive That Changed the World*. New York: Basic Books.

Kibble, T. W. B., and F. H. Berkshire. 1996. *Classical Mechanics*. Harlow, England: Longman.

Klingenberg, G. 2004. "Gun muzzle blast and flash." *Propellants, Explosives, Pyrotechnics* 14:57–68.

Landels, J. G. 1978. *Engineering in the Ancient World*. London: Constable.

Lauber, L. E. 2005. *Bowhunter's Guide to Accurate Shooting*. Chanhassen, MN: Creative Publishing.

Lee, I., and B. Kosko. 2005. "Modeling bruises in soft body armor with an adaptive fuzzy system." *IEEE Transactions on Systems, Man, and Cybernetics*. Pt. B, *Cybernetics* 35:1374–90.

Loyd, A. 1999. *My War Gone By, I Miss It So*. London: Penguin.

Maitland-Addison, J. 1918. "The long-range guns." *Field Artillery Journal* 8: 321–41.

Marshall, S. L. A. 1987. *World War I*. Boston: Houghton Mifflin.

Massey, B. S. 1989. *Mechanics of Fluids*. London: Chapman and Hall.

McCormick, R. R. 1933. "French medieval artillery." *Field Artillery Journal*, March–April, 124–28.

McPherson, J. M. 1988. *Battle Cry of Freedom: The American Civil War*. London: Penguin.

Montgomery, B. 1972. *A Concise History of Warfare*. London: Collins.

Montgomery, J. S., and E. S. Chin. 2004. "Protecting the future force: A new generation of metallic armors leads the way." *AMPTIAC Quarterly* 8:15–20.

Nennstiel, R. 1996. "How do bullets fly?" *AFTE Journal* 28: S104–43.

    Nennstiel has provided a downloadable online resource for those who are interested in the mathematical details: www.nennstiel-ruprecht.de/bullfly/index.htm.

Nicolson, A. 2005. *Seize the Fire*. New York: Harper Collins.

Norris, J., and J. Marchington. 2003. *Anti-Tank Weapons*. London: Brassey's.

Parker, G., ed. 2000. *The Cambridge Illustrated History of Warfare*. Cambridge: Cambridge University Press.

Patrick, U. W. 1989. "Handgun wounding factors and effectiveness." FBI Academy Firearms Training Unit report, July 14. Available at www.fire armstactical.com/hwfe.htm.

Pauly, R. 2004. *Firearms: The Life Story of a Technology*. Baltimore: Johns Hopkins University Press.

Payne, C. M. 2006. *Principles of Naval Weapons Systems*. Annapolis, MD: U.S. Naval Institute Press.

Phillippes, H. 1669. *A Mathematical Manual*.1st ed. London.

Ross, S. H. 2002. *Strategic Bombing by the United States in World War II: The Myths and the Facts*. Jefferson, NC: McFarland.

Rumford, Count (Benjamin Thompson). 1876. *The Complete Works of Count Rumford*. Vol. 2. London: Macmillan.

Sabin, P., H. van Wees, and M. Whitby. 2007. *Greece, the Hellenistic World, and the Rise of Rome*. Vol. 1 of *The Cambridge History of Greek and Roman Warfare*. Cambridge: Cambridge University Press.

Settles, G. S., et al. 2004. "Full-scale high-speed schlieren imaging of explosions and gunshots." SPIE Paper 5580-174. In *Proceedings of the 26th International Conference on High-Speed Photography and Photonics*, ed. D. L. Paisley. Alexandria, VA, Sept. 20–24. Also available at www.mne.psu.edu/psgdl/fullscalegunshot.pdf.

Sidnell, P. 2007. *Warhorse: Cavalry in Ancient Warfare*. London: Continuum Books.

Smith, R. D., and K. DeVries. 2005. *The Artillery of the Dukes of Burgundy, 1363–1477*. Rochester, NY: Boydell and Brewer.

Thomson, W.T. 1986. *Introduction to Space Dynamics*. New York: Dover.

U.S. Department of Defense. 1996. *Fire Control Systems—General*. Department of Defense Handbook. MIL-HDBK-799(AR). April. Available at www.every spec.com/MIL-HDBK/MIL-HDBK+(0700+-+0799)/MIL_HDBK_799 _2141.

USMC (U.S. Marine Corps). 2007. USMC instructional video: m82a1 sasr sniper rifle. October 15. http://mefeedia.com/entry/usmc-instructional-video-m82a1-sasr-sniper-rifle/14031294/.

> The Marines explain their .50 caliber sniper rifle.

U.S. Navy, Department of Ordnance and Gunnery. 1955. *Naval Ordnance and Gunnery.* Vol. 2, *Fire Control.* NavPers 10797,10798-A. Available at www.eugeneleeslover.com/USNAVY/CHAPTER-15-A.html.

Van Riper, B. 2007. *Rockets and Missiles: The Life Story of a Technology.* Baltimore, MD: Johns Hopkins University Press.

Vegetius (Flavius Vegetius Renatus). 390. *De Re Militari* [The Military Institutions of the Romans]. Available at www.pvv.ntnu.no/~madsb/home/war/vegetius.

Webster, G. 1980. *The Roman Invasion of Britain.* London: Book Club Associates.

Weinstock, R. 2002. "Cannon's recoil against a tree: How the range increases." *The Physics Teacher* 40:210–12.

Whiting, R. A. 1971. "The chemical and ballistic properties of black powder." *Explosives and Pyrotechnics* 4, nos. 1–3. Available at www.dtic.mil/cgi-bin/GetTRDoc?AD=ADA363142&Location=U2&doc=GetTRDoc.pdf.

Wolff, L. 1980. *In Flanders Fields: The 1917 Campaign.* Alexandria, VA: Time-Life Books.

YouTube. 2007a. Firing a Civil War rifle—with commands. Video. December 23. www.youtube.com/watch?v=doqgPsmT7tc&feature=related.

> A video showing a percussion cap rifle musket. These weapons were used in the American Civil War and the Crimean War.

———. 2007b. Musket demonstration. Video. September 22. www.youtube.com/watch?v=BXU0Vbgv9sE&NR=1.

> An educational demonstration of Civil War caplock rifle-musket operation.

———. 2007c. The archer's paradox recorded with high speed video. January 2. www.youtube.com/watch?v=aNI9BG87qcI.

———. 2007d. The matchlock musket. Video. November 21. www.youtube.com/watch?v=KUE4dzFPFJU.

> An insightful demonstration of English Civil War–era (mid-1600s) matchlock musket operation.

———. 2007e. 0.577 T-Rex Rifle Recoil. August 13. www.youtube.com/watch?v=F4juEIK_zRM.

> Video showing shooters trying to cope with the recoil of a 0.577 caliber T-Rex heavy hunting rifle.

———. 2008a. A demonstration of a British Brown Bess musket. Video. April 23. www.youtube.com/watch?v=lrRSy55XJNA&feature=related. Also see http://guns.wikia.com/wiki/User:Brain40, which has an animated image of a musketeer with a matchlock musket.

———. 2008b. Archer's paradox—super slow motion video. October 4. www .youtube.com/watch?v=WzWrcpzuAp8.

Zukas, J., et al. 1982. *Impact Dynamics*. Hoboken, NJ: Wiley.

Index